CONSTRUCTIVISM IN SCIENCE EDUCATION

Constructivism in Science Education
A Philosophical Examination

edited by

MICHAEL R. MATTHEWS

*School of Education Studies,
The University of New South Wales*

KLUWER ACADEMIC PUBLISHERS

DORDRECHT / BOSTON / LONDON

Library of Congress Cataloging-in-Publication Data

```
Constructivism in science education : a philosophical examination /
   edited by Michael R. Matthews.
       p.   cm.
   Includes bibliographical references and index.
   ISBN 0-7923-4924-5
   1. Science--Study and teaching--Philosophy.  2. Constructivism
 (Philosophy)    I. Matthews, Michael R.
 Q181.C612  1998
 507'.1--dc21                                              97-47347
```

ISBN 0-7923-4924-5 (PB)
ISBN 0-7923-5033-2 (HB)

Published by Kluwer Academic Publishers,
P.O. Box 17, 3300 AA Dordrecht, The Netherlands.

Sold and distributed in the U.S.A. and Canada
by Kluwer Academic Publishers,
101 Philip Drive, Norwell, MA 02061, U.S.A.

In all other countries, sold and distributed
by Kluwer Academic Publishers,
P.O. Box 322, 3300 AH Dordrecht, The Netherlands.

Printed on acid-free paper

All Rights Reserved
© 1998 Kluwer Academic Publishers
No part of the material protected by this copyright notice may be reproduced or
utilized in any form or by any means, electronic or mechanical,
including photocopying, recording or by any information storage and
retrieval system, without written permission from the copyright owner.

Printed in the Netherlands

to

Wallis Suchting (1931-1997)

friend and teacher

TABLE OF CONTENTS

SOURCES viii
PREFACE ix

MICHAEL R. MATTHEWS / Introductory Comments on Philosophy and
    Constructivism in Science Education 1

ERNST VON GLASERSFELD / Cognition, Construction of Knowledge, and
    Teaching 11

ROBERT NOLA / Constructivism in Science and in Science Education:
    A Philosophical Critique 31

WALLIS A. SUCHTING / Constructivism Deconstructed 61

ERNST VON GLASERSFELD / Constructivism Reconstructed: A Reply to
    Suchting 93

MARK H. BICKHARD / Constructivisms and Relativisms: A Shopper's Guide 99

RICHARD E. GRANDY / Constructivisms and Objectivity: Disentangling
    Metaphysics from Pedagogy 113

HELGE KRAGH / Social Constructivism, the Gospel of Science, and the
    Teaching of Physics 125

DENIS C. PHILLIPS / Coming to Terms with Radical Social Constructivisms 139

PETER SLEZAK / Sociology of Scientific Knowledge and Science Education 159

WALLIS SUCHTING / Reflections on Peter Slezak and the 'Sociology of
    Scientific Knowledge' 189

MICHAEL R. MATTHEWS / Educational Constructivism and Philosophy:
    Some References 217

NOTES ON THE CONTRIBUTORS 225

NAME INDEX 227
SUBJECT INDEX 231

SOURCES

Matthews, M.R.: 1997, 'Introductory Comments on Philosophy and Constructivism in Science Education', *Science & Education* **6**(1-2), 5-14.

Glasersfeld, E. von: 1989, 'Cognition, Construction of Knowledge and Teaching', *Synthese* **80**(1), 121-140.

Nola, R.: 1997, 'Constructivism in Science and in Science Education: A Philosophical Critique', *Science & Education* **6**(1-2), 55-83.

Suchting, W.A.: 1992, 'Constructivism Deconstructed', *Science & Education* **1**(3), 223-254.

Glasersfeld, E. von: 1992, 'Constructivism Reconstructed: A Reply to Suchting', *Science & Education* **1**(4), 379-384.

Bickhard, M.H.: 1997, 'Constructivism and Relativisms: A Shopper's Guide', *Science & Education* **6**(1-2), 29-42.

Grandy, R.E.: 1997, 'Constructivism and Objectivity: Disentangling Metaphysics from Pedagogy', *Science & Education* **6**(1-2), 43-53.

Kragh, H.: 1998, 'Social Constructivism, the Gospel of Science and the Teaching of Physics', *Science & Education* **7**(3)

Phillips, D.C.: 1997, 'Coming to Terms with Radical Social Constructivisms', *Science & Education* **6**(1-2), 85-104.

Slezak, P.: 1994a, 'Sociology of Science and Science Education: Part I', *Science & Education* **3**(3), 265-294.

Suchting, W.A.: 1997, 'Reflections on Peter Slezak and the "Sociology of Scientific Knowledge"', *Science & Education* **6**(1-2), 151-195. Note: pps. 156-170 of this 1997 article are not included in the paper published in this anthology.

PREFACE

All of these papers, with the exception of one, have been published in the journal *Science & Education*, a bimonthly research journal associated with the International History, Philosophy and Science Teaching Group. The exception, Ernst von Glasersfeld's 'Cognition, Construction of Knowledge, and Teaching' was published in a special issue of the journal *Synthese* which was prepared for the International Group's first conference at Florida State University in 1989.

The International Group aims to improve school and university science education by making it more informed by the history, philosophy, and sociology of science. The Group has a particular interest in bringing history, philosophy and sociology of science into teacher-education programmes, and in applying these disciplines to theoretical and pedagogical issues in science education.

This book brings together the work of philosophers, historians of science, cognitive psychologists and educators in an effort to situate the epistemological claims of constructivism in the history of philosophical ideas, and to advance the understanding and evaluation of constructivist claims in science education.

The *Introduction* mentions some of the philosophical issues raised by constructivism, and by the spread of constructivism from its learning-theory core into epistemology, educational theory and ethics. *Ernst von Glasersfeld*, formerly professor of psychology at the University of Georgia and now a member of the Institute for Scientific Reasoning at the University of Massachusetts, Amherst, argues for a radical constructivist view of scientific and personal knowledge, a view which traces its lineage to Vico and which is instrumentalist and anti-realist. *Robert Nola*, an associate professor of philosophy at the University of Auckland draws on the history of epistemology to provide a philosophical critique of some constructivist writing in education. *Wallis Suchting*, formerly an associate professor of philosophy at the University of Sydney, provides a detailed and critical commentary on the radical constructivist position of Ernst von Glasersfeld. Among other things he contends that empiricism, in any form, is an inadequate epistemology for science. *Ernst von Glasersfeld* replies to this criticism by defending instrumentalism and its notion of viability as the criterion of knowledge in science and personal life. *Mark Bickhard*, a professor of cognitive science and philosophy at Lehigh University, elaborates a taxonomy of constructivism and, as a constructivist, suggests that the position is not tied to relativism or to idealism, both of which he claims need to be rejected. *Richard Grandy*, a professor of philosophy at Rice University, addresses the important question of disentangling the philosophy and metaphysics of constructivism from its pedagogical practice. He says that the claims of cognitive constructivism need to be distinguished from those of metaphysical constructivism, and that cognitive constructivism has strong empirical support and suggests important new directions for the conduct of science education. *Helge Kragh*, a professor of the history of science at the University of Aarhus, poses a number of difficulties for social constructivist analyses of the history of science, difficulties that do not confront realist interpretations. He maintains that the rejection of social constructivism does not entail the rejection of social or historical studies of science, and urges the inclusion of these in the science curriculum. *Denis Phillips*, a professor of education and of philosophy at Stanford University, writes critically on social constructivism, saying that strong and exciting, but untenable, claims are advanced, and then frequently qualified to the level of tritenesss. *Peter Slezak*, a senior lecturer in cognitive science and philosophy of science at the University of New South Wales, recognises that the core claims of the Edinburgh strong programme in the sociology of scientific knowledge, if true, have a major impact on the traditional rationale for science education in schools, but argues that these claims are flawed. *Wallis Suchting*, in a second contribution, provides a detailed analysis of Peter Slezak's argument against social constructivism, and defends the core of David Bloor's social constructivist account against Slezak's criticisms. Suchting sketches an epistemological position that avoids the errors of rationalism on the one hand, and social constructivism on the other, while embracing their respective strengths. The book concludes with a hundred-item *Bibliography* of articles and books that have addressed philosophical issues in educational constructivism.

Constructivism is a very influential theory of science education. In the past three decades there have been over five thousand scholarly articles published on aspects of constructivism and education, countless conference presentations have been made, there have been scores of books on the subject, and at least one book series devoted to exploring the ramifications for education of constructivist learning theory and epistemology. The Preface of a 1993 American Association for the Advancement of Science anthology on *The Practice of Constructivism in Science Teaching* says that 'there is widespread acceptance of constructivism' and that constructivism represents a 'paradigm change' in science education. Constructivism is associated with a variety of philosophical positions in epistemology, ontology, politics and ethics. Unfortunately these philosophical aspects of constructivism are frequently taken for granted, or asserted without argument or awareness of the tradition or depth of debate that has occurred around them. For instance two leading constructivists have recently written that 'the authority for truth lies within each of us'. This claim, which goes back at least to Protagoras in the 4th century BC, if true, is truly breath-taking in its cultural and epistemological ramifications. But the claim is made without any argument, or any consideration of its obvious flaws.

As with most *isms*, constructivism has suffered its share of confused interpretations. And, to complicate matters, there are a variety of constructivisms: there are constructivist theories of learning – originally the personal constructivism of Piaget and the social constructivism of Vygotsky; there are constructivist theories of teaching that in many ways echo ancient Socratic injunctions and the more recent appeals of progressivist pedagogy; and there are constructivist theories of knowledge, or constructivist epistemologies. Clearly these varied interpretations and positions need to be untangled, and then evaluated. These are the usual tasks of philosophical analysis.

Educational constructivism – originally learning theories and theories of pedagogy – has got caught up in larger epistemological and philosophical disputes about the nature of science and about human knowledge. In part these are the so-called 'science wars' prompted by the claims of many social constructivists in philosophy of science, many of the Edinburgh 'Strong Programme' in the sociology of science, and many cultural historians of science. The flavour of these claims can be seen in a recent piece on the Scientific Revolution by a cultural historian who writes: 'Historians today feel the need to defenestrate science, or at least take it off its pedestal. Knowledge is no transcendental force for progress. Historically understood it is local, it is plural, it embodies interests, it mobilizes the claims of groups and classes, and, above all, it is recruited, willy-nilly, on all sides in wars of truth'.

Needless to say scientists, and many philosophers of a realist persuasion, have not taken these post-modernist claims lying down. The debate, joined by scientists Lewis Wolpert, Paul Gross, Norman Levitt and Alan Cromer, and philosophers such as Michael Devitt, Larry Laudan, Ernan McMullin, Ian Hacking, Mario Bunge and David Stove has been, to put it mildly, vigorous. The widely reported hoax (a spoof postmodernist attack on science) perpetuated on an academic journal by the physicist David Sokal is indicative of the nature and extent of the debate. So also is the need felt by the New York Academy of Sciences to host a symposium titled *A Defense of Science and Rationality* (published in 1997 by Johns Hopkins University Press).

It is understandable that educators have got caught up in these science wars. An item in, indeed a goal of, nearly all science curricula is 'Learning About the Nature of Science'. Teachers are expected to teach this topic, and students are expected to learn something about it. Indeed many think that knowledge of the nature of science is the most important legacy of a science education. As this topic is at the very heart of the science wars, educators who pay attention to wider intellectual issues cannot but get caught up by the debates that are raging.

Epistemological questions do not go away even if the large-scale science wars are ignored. Teachers are concerned with children gaining knowledge, thus they have to be attentive, as Socrates was two-and-a-half thousand years ago, to what knowledge is. This involves teachers in epistemology, an involvement that is increasingly recognised in the science education community, but one for which teachers are inadequately prepared by standard programmes of science teacher education. Whatever the responsibility of classroom teachers to be acquainted with core philosophical questions, the responsibilities of teacher educators is far greater. The contributions in this volume will hopefully be of assistance to teacher educators.

Clarification of language, and of concepts, has been one of the core features of philosophy since before Socrates asked at the beginning of his dialogues 'What do you mean by .....?' Contributors to this volume acknowledge the importance of such philosophical analysis for sensible and informed resolution of complex intellectual issues, and thus for educational policies, curricula and programmes. This analysis requires a willingness to become acquainted with elements of the history of philosophical and educational ideas.

In science education research and debate there is a need for patience and attention to meaning – in other words philosophy. Too often, crucial concepts, and claims in arguments, are ambiguous or confused. For example, a recent article has claimed that 'Science Studies has shown that science is socially determined'. Clearly a lot hinges on what is meant by 'socially determined' (or 'socially constructed' as it is more frequently expressed) but the article does not elaborate this linchpin claim. In one sense – science requires social resources for its conduct, its research activities reflect social priorities, and science is a communal endeavour – the claim is a truism. But in another sense – that science knowledge claims are vindicated by social agreement, not by how the world is – the claim is highly contentious. A great deal of the argument in the article is flawed by the author's failure to specify what sense of 'social determination' is meant. The author engages in the common practice of 'soft-focus' writing. This assists relaxed reading, but it inhibits understanding and the appraisal of ideas.

There is also a tendency to use philosophical terms without due attention to their meaning, indeed to use them recklessly. Terms such as 'positivist', 'realist', 'logical positivist', 'empiricist', 'inductivist' 'objectivist' etc. are used very loosely. Such loose usage does not advance understanding. Indeed in science education literature 'positivist' and 'realist' are frequently synonyms, while in the history of philosophy they have denoted contrary epistemological positions – the former maintaining that science can only have knowledge of observables and that theoretical terms are merely 'aids to thought', the latter denying this position.

There is also an unfortunate propensity for intellectual laziness which is manifest in 'piggy-back' argument, or argument by citation. It is common to see claims such as 'Since Kuhn so and so', 'Latour established that such and such', or 'Science studies has shown the following'. The 'so and so', 'such and such', and 'following' are usually very contentious and much debated claims in the professional historical and philosophical communities. Notoriously in the case of Kuhn, there is considerable disagreement about what he did and did not mean to establish, and he was at pains to separate himself from many 'Kuhnians'. Science education needs to go back to philosophical sources, and to pay attention to detail. It is easy to substitute citation for argument and analysis, but this easy option does not advance the resolution of complex and important matters. And it is a poor intellectual model for students. The step from the acceptance of sloppy thinking, to thinking in slogans, is very small and, unfortunately, much trodden.

One way to reduce these deficiencies in science education debate is to have history and philosophy of science courses included in preservice and graduate programmes. This is one of the goals that the journal *Science & Education* tries to promote. In the first number of the journal it was stated that:

A particular concern of *Science & Education* is the improvement of science and mathematics teacher education programmes. ... Teachers have a particular responsibility to understand their subject matter, ... its place in the intellectual scheme of things, how it has evolved, its epistemology, its connections with ethical, political and religious questions. They need not just a first-order understanding of the subject, that is its content, but a second-order understanding.

There is no doubt that constructivism has presented great challenges to science teachers who seek to 'understand their subject matter and its place in the intellectual scheme of things'. I hope that the patient reading of these contributions will assist teachers and educators in this task, and contribute to a more sophisticated discussion of these matters in the science education community.

Michael R. Matthews
School of Education Studies
University of New South Wales
Sydney, 2052
AUSTRALIA

Introductory Comments on Philosophy and Constructivism in Science Education

MICHAEL R. MATTHEWS

School of Education Studies, University of New South Wales, Sydney 2052, Australia

ABSTRACT: This article indicates something of the enormous influence of constructivism on contemporary science education. The article distinguishes educational constructivism (that has its origins in theories of children's learning), from constructivism in the philosophy of science (usually associated with instrumentalist views of scientific theory), and from constructivism in the sociology of science (of which the Edinburgh Strong Programme in the sociology of scientific knowledge is the best known example). It notes the expansion of educational constructivism from initial considerations of how children come to learn, to views about epistemology, educational theory, ethics, and the cognitive claims of science. From the learning-theory beginnings of constructivism, and at each stage of its growth, philosophical questions arise that deserve the attention of educators. Among other things, the article identifies some theoretical problems concerning constructivist teaching of the content of science.

Constructivism is a major influence in contemporary science and mathematics education. And in its post-modernist or 'deconstructivist' version, it is a significant influence in literary, artistic, historical, social science and even theological education. Constructivism, in all its varieties, has been the subject of heated debate. The debate is not simply about the adequacy of a particular learning theory, or the cogency of an epistemological position, or the fruitfulness of a hermeneutical principle. Something more is at stake. This is suggested by the frequency with which constructivists link their position to ideas of empowerment and emancipation – ideas that clearly outrun mere learning theory or epistemology. Many commentators have remarked on how frequently constructivists stake out the moral high ground for their position (see especially Kilpatrick 1987). But their occupancy is not without dispute. Paul Gross and Norman Levitt's book *Higher Superstition* (Gross & Levitt 1994) has given wide popular exposure to the anti-constructivist cause. Michael Devitt has written that:

> I have a candidate for *the* most dangerous contemporary intellectual tendency, it is ... constructivism. Constructivism is a combination of two Kantian ideas with twentieth-century relativism. The two Kantian ideas are, first, that we make the known world by imposing concepts, and, second, that the independent world is (at most) a mere 'thing-in-itself' forever beyond our ken. ... [considering] its role in France, in the social sciences, in literature departments, and in some largely well-meaning, but confused, political movements [it] has led to a veritable epidemic of 'worldmaking'. Constructivism attacks the immune system that saves us from silliness. (Devitt 1991, p. ix).

Debate not withstanding, constructivism is influential in Western education. A former president of the US National Association for Research in Science Teaching (NARST) has said that 'A unification of thinking, research, curriculum development, and teacher education appears to now be occurring under the theme of constructivism... there is a lack of polarised debate' (Yeany 1991, p. 1 – for references see Bibliography at end of issue). The American Association for the Advancement of Science has published a nineteen-chapter book titled *The Practice of Constructivism in Science Education*, the Preface of which states that 'there is widespread acceptance of constructivism', and that 'constructivism has become increasingly popular... in the past 10 years', indeed it represents a 'paradigm change' in science education (Tobin 1993 – see Book Notes section of this issue). A 1990 bibliography produced at Leeds University, a major centre of constructivist research, listed over 1,000 works (Carmichael et al. 1990). Reinders Duit, at the Institute for Science Education in Kiel, has been performing the Herculean task of keeping up-to-date with research in this field, and his estimate is that there are currently 2,500 constructivist-inspired scholarly research articles in education journals (Duit 1993). The constructivist research programmes at Monash University, Waikato University, Florida State University, and Leeds University – to name perhaps the major centres – have been energetic and productive.

Although there have been some critics of constructivism (Suchting 1992, Matthews 1993, Phillips 1995, Osborne 1996), and some urging caution in its adoption (Millar 1989, Solomon 1993), few would dispute Peter Fensham's claim that 'The most conspicuous psychological influence on curriculum thinking in science since 1980 has been the constructivist view of learning' (Fensham 1992, p. 801).

THE TRANSFORMATION OF CONSTRUCTIVISM

It is important to notice that Fensham speaks of 'psychological influence', and the constructivist 'view of learning'. This is the original core of constructivism – a psychological theory about how beliefs are developed, not what makes beliefs true or what counts as scientific knowledge. From this core, constructivism has expanded to incorporate views about epistemology, teaching, curriculum, educational theory, ethics, ontology, and metaphysics.

But even at its learning-theory core epistemology is involved. Jean Piaget well recognised that any decent learning theory involves epistemological considerations – he called his own research programme 'Genetic Epistemology', and one of his most influential books is titled *Psychology and Epistemology: Towards a Theory of Knowledge* (Piaget 1972). More recently Kenneth Strike and George Posner have said that their oft-cited 1982 paper – 'Accommodation of a Scientific Conception: Toward a Theory of Conceptual Change' (Posner et al. 1982) – is 'largely an epistemological theory, not a psychological theory... it is rooted in a conception of the kinds of things that count as *good* reasons' (Strike & Posner 1992, p. 150). Thus from the core outwards, constructivism necessarily involves considerations of philosophy – something for which, unfortunately, science teachers are often ill-prepared by their training.

There are of course many varieties of constructivism (see for instance Good et al. (1993), Phillips (1995) and Geelan (this issue)). For the sake of convenience these may be divided into educational constructivism, philosophical constructivism, and sociological constructivism. The first variety itself divides into personal and social constructivism. The second variety has its immediate origins in Thomas Kuhn's work, and is most robustly represented by Bas van Fraasen, a recent President of the US Philosophy of Science Association. This philosophical constructivism has its roots in Berkeley's philosophy of science, and further back in instrumentalist philosophy of Ancient Greece. This tradition has been, since Aristotle, opposed by realists in the philosophy of science (see Matthews 1994, chap. 8). Sociological constructivism is identified with the Edinburgh 'Strong Programme' and their research on the Sociology of Scientific Knowledge (SSK). In this tradition the growth of science, and changes in its theories and philosophical commitments, is interpreted in terms of changing social conditions and interests; and the explanatory power of cognitive content and rational reasoning is discounted. That is, that something is true and reasonable is not thought, by adherents to the Edinburgh Programme, to constitute an explanation of why it is believed.

PHILOSOPHY AND EDUCATIONAL CONSTRUCTIVISM

Educational constructivism draws upon the other constructivist – philosophical and sociological – traditions, but it has its own autonomous roots and history. Educational constructivism of the personal variety stresses the individual creation of knowledge and construction of concepts. This stream has its origins in Piaget's Kantian-inspired theories of cognitive development, and Ernst von Glasersfeld is perhaps its best known representative (see Hardy & Taylor, this issue). Educational constructivism of the social variety stresses the importance of the group (be it the immediate classroom or the wider culture) for the development and validation of ideas. This has its origins in Vygotsky's work in linguistics and language acquisition, and is seen, for instance, in the later publications of Rosalind Driver.

A working distinction can be drawn between constructivist theory and constructivist pedagogy. The need for this distinction was brought home to me some years ago at an international science education conference when I was raising epistemological questions with one leading constructivist, who replied: "Michael, what I am interested in is good teaching, all the other stuff [epistemology, cognitive psychology] is peripheral". There is of course, a variety of constructivist postions on this theory/practice matter. Many think constructivist epistemology and cognitive psychology is germane to the pedagogical prescriptions; others like the conference presenter think that the theory is much less important than the practice; others concentrate solely on pedagogy, and improved classroom practices, and simply use the label 'constructivist' to refer to anything which is pupil-centered, engaging, questioning, and progressive. For the latter, the details of epistemology and cognitive psychology are unimportant, and not worth disputing about. They might be called, if labels are required, 'pedagogical constructivists'.

The collection of papers in this double-issue of *Science & Education* are all focussed on philosophical, particularly epistemological, aspects of educational constructivism, or those parts of the other constructivist traditions that have been utilised in the educational literature. Guides to the debate, and literature, about constructivism in the philosophy of science can be found in Leplin (1984), Churchland & Hooker (1985), and Brown (1994). Comparable guides to constructivism in the sociology of scientific knowledge can be found in Brown (1984) and McMullin (1988, 1992).

The guiding idea behind this journal issue is that the epistemological claims of educational constructivism are important, and that they deserve further delineation and investigation.

The New Zealand Example

Unfortunately matters of deep philosophical importance over which there have been centuries of debate, too frequently appear almost as throwaway lines in science education writing. When they are elaborated, the elaboration is often slight, being little more than the citing of names, or claims that 'since Kuhn such and such', or 'following Kant so and so', or 'Latour has shown that something or other'. There have of course been deeper analyses, many of which are discussed by contributors, and cited in this issue's bibliography. But overall the theoretical, pedagogical and curriculum proposals of educational constructivism are like an inverted pyramid: they rest on a tiny base. It is in everyone's interest that this base be made more substantial, and be well scrutinised.

For example, the very influential and much-reprinted book of Roger Osborne and Peter Freyberg – *Learning in Science: The Implications of Children's Science* (Osborne & Freyberg, 1985) opens with the statement that:

> Young children and scientists have much in common. Both are interested in a wide variety of objects and events in the world around them. Both are interested in, and attempt to make sense of, how and why things behave as they do. (Osborne & Freyberg 1985, p. 1)

This idea – that science is about making sense of the world, rather than finding out about the world – has been much debated in the history of philosophy. But the book does not elaborate the debate or defend its position.

The book concludes with a statement of the *sine qua none* of constructivism – the non-transferability of knowledge thesis:

> Knowledge is acquired not by the internalisation of some outside given but is constructed from within. (Osborne & Freyberg 1985, p. 82)

Again there is no elaboration or defense of this, to put it mildly, contentious position. The alternative position in learning theory is one that maintains:

> if you want somebody to know something, you teach it to them . . . if you want somebody to know something and retain it for a long period of time, then you have them practice it. (Geary 1995, p. 33)

These casual, almost throw-away epistemological positions have cast a long shadow in New Zealand education. For instance, the New Zealand National Curriculum in Science (1993) labels the chief learning areas: 'Making Sense of the Living World' (instead of Biology, or Finding Out About the Living World), and 'Making Sense of Planet Earth and Beyond' (instead of Astronomy, or Finding Out About the Solar System). And a 1990 draft syllabus identifies the 'Role of the Teacher' as being: 'Helping students learn how to learn; being a learner too; ensuring equity for all students; creating a friendly, supportive learning environment; providing learning opportunities; listening to students; using the students' ideas, experiences, and interests; challenging sensitively the ideas of students; providing resources to help students learn; ensuring students communicate in a variety of modes; identifying and nurturing the scientific talent and interests of all students – provided that teachers are aware of the effectiveness of an open science programme which allows students to realise their own potential at their own pace; and finally, contributing to the planning of the school science programme'. This list has everything except knowing the subject matter to be taught, and being able to teach it in a clear, engaging and understandable manner.

There is a not-too-subtle difference between the constructivist formulation 'making sense', and the realist formulation 'finding out'. The former has no epistemological or referential bite; the latter has both. Things can make perfect sense without being true; and making still more sense does not imply any increase in truth content.

At this stage the point of the New Zealand example (elaborated in Matthews 1995) is only to illustrate the claim that frequently in science education far-reaching philosophical claims are made without the intellectual support that their importance warrants. Many rush into philosophical positions, where fearing to tread is a wiser policy.

ETHICS, POLITICS AND IDEOLOGY IN CONSTRUCTIVIST THOUGHT

Thus far, constructivist theory has been spoken about in its *epistemological* guise, and three varieties – educational, philosophy of science, and sociological – have been mentioned. But epistemology is not the only area into which constructivist theory has expanded.

Constructivism also has an *ethical* dimension. A recent paper says 'There is also a sense in which constructivism implies caring – caring for ideas, personal theories, self image, human development, professional esteem, people – it is not a take-it-or-leave-it epistemology' (Watts 1994, p. 52). This ethical dimension is manifest in the frequency with which notions of emancipation and empowerment occur in constructivist writing. Constructivism is thought to be a morally superior position to its rivals in learning theory and pedagogy (see Hardy & Taylor, this issue).

There is also a *political* dimension to much constructivist writing. Two constructivist writers say that they are 'committed to the philosophy and principles of composite grades and mixed-ability groupings' (Brass & Duke 1994, p. 100). Another writer has identified the Progressive Education tradition as constructivist, and the British Plowden Report of the mid-1960s as the embodiment of constructivist school organisation (Hawkins 1994).

A number of constructivists align themselves with the Critical Theory of Michael Apple, Henry Giroux and Stanley Aronowitz. One New Zealand commentator says that 'There are many parallels between the literature on the development of critical pedagogy [and] the literature on constructivist learning' (Gilbert 1993, p. 35). This is because 'Critical theorists question the value of such concepts as individualism, efficiency, rationality and objectivity, and the forms of curriculum and pedagogy that have developed from these concepts' (Gilbert 1993, p. 20).

One commentator sees constructivism as a component of cultural ideology, saying that:

> In sum, constructivism is largely a reflection of current American cultural beliefs and, as such, involves the development of instructional techniques that attempt to make the acquisition of complex mathematical skills an enjoyable social enterprise that will be pursued on the basis of individual interest and choice. (Geary 1995, p. 32)

For some, constructivism has gone beyond learning and educational theory, into the realms of metaphysics and worldviews. Ken Tobin, a past president of the US National Association for Research in Science Teaching, has said:

> To become a constructivist is to use constructivism as a referent for thoughts and actions. That is to say when thinking or acting, beliefs associated with constructivism assume a higher value than other beliefs. For a variety of reasons the process is not easy. (Tobin 1991, p. 1)

What needs to be recognised is that many things are being said in the name of constructivism that transend its original domain of learning theory. It is frequently assumed that constructivist learning theory has these flow-on effects into epistemology, educational theory, political theory, ideology, and metaphysics. For many, constructivism presents itself as a package deal: if you buy the learning theory, then you buy the epistemology and everything else. Conversely, if you are in favour of educational reform, and the emancipation and liberation of human beings, then you must be a constructivist in learning theory and pedagogy.

Clearly we need to disentangle, and to examine carefully, these different elements. The contributors to this special issue largely focus on the epistemological aspects of constructivist theory. But this is not to say that the ethical, political and educational – in the sense of educational theory – aspects of constructivism are unimportant. They certainly are important, and also need to be thoroughly examined. It is now basically accepted that when science educators speak of epistemological matters they need to be, to some extent, familiar with the tradition of epistemological debate in philosophy. So educators who speak of ethics, emancipation, empowerment and social justice should also be familiar with the long tradition of philosophical debate about ethics, politics, educational theory, and social theory. Without this familiarity (something that courses in philosophy can provide), the risk is that educators merely repeat popular nostrums, or voice politically correct fashions.

Constructivism has done a service to science education by alerting teachers to the function of prior learning and extant concepts in the process of learning new material, by stressing the importance of understanding as a goal of science instruction, by fostering pupil engagement in lessons, and other such progressive matters. But liberal educationalists can rightly say that these are pedagogical commonplaces, the recognition of which goes back at least to Socrates. It is clear that the best of constructivist pedagogy can be had without constructivist epistemology – Socrates, Montaigne, Locke, Mill, and Russell are just some who have conjoined engaging, constructivist-like pedagogy with non-constructivist epistemology.

Constructivism has also done a service by making educators aware of the human dimension of science: its fallibility, its connection to culture and interests, the place of convention in scientific theory, the historicity of concepts, the contested nature of theories, and much else. But again realist philosophers can rightly maintain that constructivism does not have a monopoly on these insights. They can be found in the work of thinkers as diverse as Mach, Duhem, Bachelard, Popper, and Polanyi.

CONSTRUCTIVIST TEACHING OF THE CONTENT OF SCIENCE

One response to criticism of constructivist theory is to say that constructivist pedagogy is valuable and should be encouraged, even if the theory is debatable. This position is understandable, but it rests on a moot point: How efficacious is constructivist pedagogy in teaching science?

One prominent constructivist, Richard White, has said 'although the research on alternative conceptions has sparked interest in content, it has not yielded clear advice about how to teach different topics (Fensham, Gunstone & White, 1994, p. 255). Given the necessity for any science programme to teach the content of science this is a serious failure.

The difficulty for constructivism posed by teaching the content of science is not just a practical one, it is a difficulty that exposes a fundamental *theoretical* problem for constructivism – if knowledge cannot be imparted, and if knowledge must be a matter of personal construction, then how can children come to knowledge of complex conceptual schemes that have taken the best minds hundreds of years to build up?

Many science educators are interested in finding out how, on constructivist principles, one teaches a body of scientific knowledge that is in large part abstract (depending on notions such as velocity, acceleration, force, gene), removed from experience (propositions about atomic structure, cellular processes, astronomic events), has no connection with prior conceptions (ideas of viruses, antibodies, molten core, evolution, electromagnetic radiation), and is alien to common-sense, and in conflict with everyday experience, expectations and concepts? Teaching a body of knowledge involves not just teaching the concepts, but also the method, and something of the methodology or theory of method. How all of this is to be taught, without teachers actually conveying something to pupils, is a moot point.

Joan Solomon, a prominent British science educator, well articulates the problem:

> Constructivism has always skirted round the actual learning of an established body of knowledge ... students will find that words are used in new and standardised ways: problems which were never even seen as being problems, are solved in a sense which needs to be learned and rehearsed. For a time all pupils may feel that they are on foreign land and no amount of recollection of their own remembered territory with shut eyes will help them to acclimatise. (Solomon, 1994, p. 16)

The constructivist research of Rosalind Driver and scholars at Leeds University illustrates the 'skirting around' to which Solomon draws attention. In a recent book the Leeds group reasonably enough maintain that:

> ... learning science involves being initiated into the culture of science. If learners are to be given access to the knowledge systems of science, the process of knowledge construction must go beyond personal empirical enquiry. Learners need to be given access not only to physical experiences but also to the concepts and models of conventional science. (Driver et al., 1994b, p. 6)

There is near unanimity on this claim – conservatives and progressivists all agree, with perhaps just discovery-learners dissenting. The claim echoes the Leeds group's oft-repeated assertion that constructivism is different from discovery learning (on this, see Miller & Driver 1987). But having made the above claim, the Leeds group go on to say that:

The challenge for teachers lies in helping learners to construct these models for themselves, to appreciate their domains of applicability and, within such domains, to use them.

One might reasonably ask whether, at this point, learning theory, or ideology, is simply getting in the way of good teaching. Why must learners construct for themselves the ideas of potential energy, mutation, linear inertia, photosynthesis, valency, and so on? Why not explain these ideas to students, and do it in such a way that they understand them? Certainly a challenge for constructivist teachers lies in helping learners construct these ideas without violating constructivist learning principles. The Leeds group recognise this, and go on to say:

> If teaching is to lead pupils towards conventional science ideas, then the teacher's intervention is essential, both through providing appropriate experiential evidence and making the theoretical ideas and conventions available to pupils. (Driver et al. 1994b, p. 6)

This is perhaps the precise point where Joan Solomon's 'skirting around' is evidenced. How can a teacher make 'the theoretical ideas and conventions available to pupils' without explaining them, without illustrating them, without showing their interconnections. In brief, without teaching them to pupils?

Constructivists addressed the problem of teaching the content of science at an international seminar held at Monash University in 1992. Its published proceedings were titled *The Content of Science: A Constructivist Approach to its Teaching and Learning* (Fensham, Gunstone & White, 1994). Rosalind Driver and colleagues made a contribution to the seminar on 'Planning and Teaching a Chemistry Topic from a Constructivist Perspective'. They had children put nails in different places and observe the rate at which they rusted. They remarked that:

> The theory that rusting is a chemical reaction between iron, oxygen and water, resulting in the formation of a new substance, is not one that students are likely to generate for themselves. (Fensham, Gunstone & White 1994, p. 206)

Indeed. After ten pages describing how the teacher tries to 'keep faith with students reasoning ... yet lead them to the intended learning goals' (p. 207), we are told that 'The process of investigating personal ideas and theories may lead students to reflect upon and question them. At the same time, it is unlikely to lead to the scientific view' (p. 218).

Quite so. But where does this leave constructivism as a putatively useful theory for science teachers?

Most science teachers realise this difficulty. They try their best to explain things clearly, to make use of metaphors, to use demonstrations and practical work to flesh out abstractions, to utilise projects and discussions for involving students in the subject matter, and so on. They realise that many, if not most, things in science are beyond the experience of students and the capabilities of school laboratories to demonstrate. The cellular, molecular and atomic realms are out of reach of school laboratories, as is most of the astronomical realm. Most of the time even things that are within reach do not work. It is a rare school experiment that is successful. For children, a great deal of science has to be taken on faith. Good teachers do their best, and try to point out why faith in science is warranted.

Some would say that the constructivist/anti-constructivist argument reduces to a mere verbal preference: constructivist teachers (of the Leeds variety) perhaps do make the concepts available in the sense of teaching them, but they prefer to talk of student construction, while traditionalists prefer to talk of transmission. Where there is a failure of match between the pupil's idea and the scientific idea, constructivists prefer to talk of imperfect construction, and traditionalists prefer to talk of failure of attention, imperfect comprehension, inadequate preparation. Provided both groups of teachers are doing the same thing, and judging the outcome by the degree to which the pupil understands the current scientific concept, then the argument could be seen as merely a verbal one over the name of a label.

But it is not just a verbal matter. A practical, but not insignificant, consideration is that science teachers are overwhelmed by challenges – pupils' lack of interest in science, teachers' inadequate knowledge of science, schools' lack of resources, society's lack of interest in education – they do not need to be further weighed down by illusory challenges, if indeed this is what the constructivist challenge amounts to. A theoretical consideration is the very justification of science in the curriculum. If Western science is truly just one among many equally warranted ways of making sense of experience, and it truly does not tell us anything about the world in which we live, then traditional arguments for compulsory school science, and 'Science for All', need to be re-thought.

CONCLUSION

The foregoing considerations serve to highlight the importance, for both constructivists and their critics, of getting clear about just what are, and are not, the epistemological commitments of constructivism. And what relationship these commitments have, if any, to classroom practice. Given the influence of constructivism on education reform, teacher education and pedagogy, these are not idle investigations.

It is hoped that this issue of *Science & Education* will advance these investigations.

ERNST VON GLASERSFELD

COGNITION, CONSTRUCTION OF KNOWLEDGE, AND TEACHING*

> The only truly ubiquitous factor in cognitive developments – be it in the history of science or in the ontogeny of mind – are of a functional, not a structural kind.
> Piaget and Garcia, 1983, p. 38

ABSTRACT. The existence of objective knowledge and the possibility of communicating it by means of language have traditionally been taken for granted by educators. Recent developments in the philosophy of science and the historical study of scientific accomplishments have deprived these presuppositions of their former plausibility. Sooner or later, this must have an effect on the teaching of science. In this paper I am presenting a brief outline of an alternative theory of knowing that takes into account the thinking organism's cognitive isolation from 'reality'. This orientation was proposed by Vico at the beginning of the 18th century, disregarded for two hundred years, and then propounded independently by Piaget as a developmentally grounded constructivist epistemology. The paper focuses specifically on the adaptive function of cognition, Piaget's scheme theory, the process of communication, and the subjective perspective on social interaction. In the concluding section it then suggests some of the consequences the shift of epistemological presuppositions might have for the practice of teaching.

During the last three decades faith in objective scientific knowledge, a faith that formerly served as the unquestioned basis for most of the teaching in schools and academia, has been disrupted by unsettling movements in the very discipline of philosophy of science. Though the roots of the subversion go back a good deal further, the trouble was brought to the awareness of a wider public by the publication of Kuhn's *The Structure of Scientific Revolutions*. There, undisguised and for everyone to read, was the explicit statement that

> ... research in parts of philosophy, psychology, linguistics, and even art history, all converge to suggest that the traditional epistemological paradigm is somehow askew. That failure to fit is also made increasingly apparent by the historical study of science.... None of these crisis-promoting subjects has yet produced a viable alternate to the traditional epistemological paradigm, but they do begin to suggest what some of that paradigm's characteristics will be. (Kuhn 1970, p. 121)

While the troubles of the "traditional epistemological paradigm" have shown no sign of subsiding in the years since Kuhn's publication, one could not honestly say that any substitute has been generally accepted. In most departments of psychology and schools of education, teaching continues as though nothing had happened and the quest for immutable objective truths were as promising as ever. For some of us, however, a different view of knowledge *has* emerged, not as a new invention but rather as the result of pursuing suggestions made by much earlier dissidents. This view differs from the old one in that it deliberately discards the notion that knowledge could or should be a representation of an observer-independent world-in-itself and replaces it with the demand that the conceptual constructs we call knowledge be *viable* in the experiential world of the knowing subject.

Ludwig Fleck, whose monograph of 1935 Kuhn acknowledged as a forerunner, wrote an earlier article in 1929 that went virtually unnoticed and that already contained much that presages what the Young Turks have been proposing in recent years:

The content of our knowledge must be considered the free creation of our culture. It resembles a traditional myth. (Fleck 1929, p. 425)

Every thinking individual, insofar as it is a member of some society, has its own reality according to which and in which it lives. (p. 426)

Not only the ways and means of problem solutions are subject to the scientific style, but also, and to an even greater extent, the choice of problems. (p. 427)

In his monograph, Fleck then cites Jakob von Uexküll (1928) as a fellow proponent of the notion of subjective realities, but criticizes him for not being radical enough. In retrospect, one might conjecture that Fleck would have agreed more fully with von Uexküll's later elaboration of the biological organisms' self-generated environments. In any case, it is this *construction* of the individual's subjective reality which, I want to suggest in this paper, should be of interest to practitioners and researchers in education and, in particular, to the teachers of science.

The notion of cognitive construction was adopted in our century by Mark Baldwin and then extensively elaborated by Jean Piaget. Piaget's *constructivist* theory of cognitive development and cognition, to which I shall return later, had, unbeknownst to him, a striking forerunner in the Neapolitan philosopher Giambattista Vico. Vico's epistemological treatise (1710) was written in Latin and remained almost unknown. Yet no present-day constructivist can afford to ignore it, because the way Vico formulated certain key ideas and the way they were briefly discussed at the time is, if anything, more relevant today than it was then.

THE ROOTS OF CONSTRUCTIVISM

The anonymous critic who, in 1711, reviewed Vico's first exposition of a thoroughly constructivist epistemology expressed a minor and a major complaint. The first – with which any modern reader might agree – was that Vico's treatise is so full of novel ideas that a summary would turn out to be almost as long as the work itself (e.g., the introduction of developmental stages and the incommensurability of ideas at different historical or individual stages, the origin of conceptual certainty as a result of abstraction and formalization, the role of language in the shaping of concepts). The reviewer's second objection, however, is more relevant to my purpose here, because it clearly brings out the problem constructivists run into, from Vico's days right down to our own.

Vico's treatise *De antiquissima Italorum sapientia* (1710), the Venetian reviewer says, is likely to give the reader "an idea and a sample of the author's metaphysics rather than to prove it". By *proof*, the 18th-century reviewer intended very much the same as so many writers seem to intend today, namely a solid demonstration that what is asserted is *true* of the real world. This conventional demand cannot be satisfied by Vico or any proponent of a radically constructivist theory of knowing: one cannot do the very thing one claims to be impossible. To request a demonstration of *Truth* from a radical constructivist shows a fundamental misunderstanding of the author's explicit intention to operate with a different conception of knowledge and of its relation to the 'real' world.

One of Vico's basic ideas was that epistemic agents can *know* nothing but the cognitive structures they themselves have put together. He expressed this in many ways, and the most striking is perhaps: "*God is the artificer of Nature, man the god of artifacts*". Over and over he stresses that "to know" means *to know how to make*. He substantiates this by saying that one knows a thing only when one can tell what components it consists of. Consequently, God alone can know the *real* world, because He knows how and of what He has created it. In contrast, the human knower can know only what the human knower has constructed.

For constructivists, therefore, the word *knowledge* refers to a commodity that is radically different from the objective representation of an observer-independent world which the mainstream of the Western philosophical tradition has been looking for.[1] Instead, *knowledge* refers to conceptual structures that epistemic agents, given the range of present experience within their tradition of thought and language, consider *viable*.

Richard Rorty, in his Introduction to *Consequences of Pragmatism*, announces this shift of focus in terms that fit the constructivist's position just as well as the pragmatist's:

He [the pragmatist] drops the notion of truth as correspondence with reality altogether, and says that modern science does not enable us to cope because it corresponds, it just enables us to cope. (Rorty 1982, p. XVII)

Constructivism *is* a form of pragmatism and shares with it the attitude towards knowledge and truth; and no less than pragmatism does it go against "the common urge to escape the vocabulary and practices of one's own time and find something ahistorical and necessary to cling to" (Rorty 1982, p. 165).

The anonymous reviewer's complaint that Vico did not *prove* his thesis, reproaches Vico for not having claimed for his 'metaphysics' (which was actually a theory of knowing) the correspondence with an ahistorical ontic world *as God might know it*. But this notion of correspondence was precisely what Vico – like the pragmatists – intended to drop.

Present-day constructivists, however, if pressed for corroboration rather than proof in the traditional sense, have an advantage over Vico. They can claim *compatibility* with scientific models that enable us to 'cope' remarkably well in specific areas of experience. For instance, one might cite the neurophysiology of the brain and quote Hebb's:

At a certain level of physiological analysis there is no reality but the firing of single neurons. (Hebb 1958, p. 461)

This is complemented by von Foerster's (1970) observation that all sensory receptors (i.e., visual, auditory, tactual, etc.) send physically indistinguishable 'responses' to the cortex and that, therefore, the 'sensory modalities' can be distinguished only by keeping track of the part of the body from which the responses come, and *not* on the basis of 'environmental features'. Such statements make clear that contemporary neurophysiological models may be compatible with a constructivist theory of knowing but can in no way be integrated with the notion of transduction of 'information' from the environment that any realist epistemology demands.

KNOWLEDGE AS AN ADAPTIVE FUNCTION

Constructivism differs from pragmatism in its predominant interest in *how* the knowledge that "enables us to cope" is arrived at. The work of Jean Piaget, the most prolific constructivist in our century, can be interpreted as one long struggle to design a model of the generation of viable knowledge. In spite of Piaget having reiterated innumerable times (cf. 1967a, pp. 210ff) that, from his perspective, cognition must be considered an *adaptive function*, most of his critics argue against him as though he were concerned with the traditional notion of knowledge as correspondence.

This misinterpretation is to some extent due to a misconception about adaptation. The technical sense of the term that Piaget intended comes from the theory of evolution. In that context, *adaptation* refers to a state of organisms or species that is characterized by their ability to survive in a given environment. Because the word is often used as a verb (e.g., this or that species *has* adapted to such and such an environment), the impression has been given that adaptation is an evolutionary activity. This is quite misleading. In phylogeny no organism can actively modify its genome and generate characteristics to suit a changed environment. According to the theory of evolution, the modification of genes is always an accident. Indeed, it is these accidental modifications that generate the variations on which natural selection can operate. And nature does *not* – as even Darwin occasionally slipped into saying (Pittendrigh 1958, p. 397) – select 'the fittest', it merely lets live those that have the characteristics necessary to cope with their environment and lets die all that have not.

This interpretation of the theory of evolution and its vocabulary is crucial for an adequate understanding of Piaget's theory of cognition. As for Vico, knowledge for Piaget is never (and can never be) a 'representation' of the real world. Instead it is the collection of conceptual structures that turn out to be adapted or, as I would say, *viable* within the knowing subject's range of experience.

In both, theory of evolution and the constructivist theory of knowing, 'viability' is tied to the concept of equilibrium. Equilibrium in evolution indicates the state of an organism or species in which the potential for survival in a given environment is genetically assured. In the sphere of cognition, though indirectly linked to survival, equilibrium refers to a state in which an epistemic agent's cognitive structures have yielded and continue to yield expected results, without bringing to the surface conceptual conflicts or contradictions. In neither case is equilibrium necessarily a static affair, like the equilibrium of a balance beam, but it can be and often is dynamic, as the equilibrium maintained by a cyclist.

To make the Piagetian definition of knowledge plausible, one must immediately take into account (which so many interpreters of Piaget seem to omit) that a human subject's experience always includes the social interaction with other cognizing subjects. This aspect of social interaction is, obviously, of fundamental importance if we want to consider education, that is, any situation in which the actions of a teacher are aimed at generating or modifying the cognitive constructions of a student. But introducing the notion of social interaction raises a problem for constructivists. If what a cognizing subject knows cannot be anything but what that subject has constructed, it is clear that, from the constructivist perspective, the *others* with whom the subject may interact socially cannot be posited as an ontological given. I shall return to this problem as well as to the constructivist approach to education; but first I want to explicate the basis of a Piagetian theory of learning.

THE CONTEXT OF SCHEME THEORY

Two of the basic concepts of Piaget's theory of cognition are *assimilation* and *accommodation*. Piaget's use of these terms is not quite the same as their common use in ordinary language. Both terms must be understood in the context of his constructivist theory of knowing. Unfortunately, this is what contemporary textbooks in developmental psychology (most of which devote at least a few pages to Piaget) often fail to do. Thus one reads, for instance:

Assimilation is the process whereby changing elements in the environment become incorporated into the structure of the organism. At the same time, the organism must accommodate its functioning to the nature of what is being assimilated. (Nash 1970, p. 360)

This is not at all what Piaget meant. One reason why assimilation is so often misunderstood is that its use as an explanatory postulate ranges from the unconscious to the deliberate. Another stems from disregarding that Piaget uses that term, as well as 'accommodation', within the framework of his theory of schemes. An example may help to clarify his position.

An infant quickly learns that a rattle it was given makes a rewarding noise when it is shaken, and this provides the infant with the ability to generate the noise at will. Piaget sees this as the "construction of a *scheme*" which, like all schemes, consists of three parts:

(1) Recognition of a certain situation (e.g., the presence of a graspable item with a rounded shape at one end);
(2) association of a specific activity with that kind of item (e.g., picking it up and shaking it);
(3) expectation of a certain result (e.g., the rewarding noise).

It is very likely that this infant, when placed in its high-chair at the dining table, will pick up and shake a graspable item that has a rounded shape at one end. We call that item a spoon and may say that the infant is *assimilating* it to its rattling scheme; but from the infant's perspective at that point, the item *is* a rattle, because what the infant perceives of it is not what an adult would consider the characteristics of a spoon but just those aspects that fit the rattling scheme.[2]

Shaking the spoon, however, does not produce the result the infant expects: the spoon does not rattle. This generates a perturbation ('disappointment'), and perturbation is one of the conditions that set the stage for cognitive change. In our example it may simply focus the infant's attention on the item in its hand, and this may lead to the perception of some aspect that will enable the infant in the future to recognize spoons as non-rattles. That development would be an *accommodation*, but obviously a rather modest one. Alternatively, given the situation at the dining table, it is not unlikely that the spoon, being vigorously shaken, will hit the table and produce a different but also very rewarding noise. This, too, will generate a perturbation (we might call it 'enchantment') which may lead to a different *accommodation*, a major one this time, that initiates the "spoon banging scheme" which most parents know only too well.

This simple illustration of scheme theory also shows that the theory involves, on the part of the observer, certain presuppositions about cognizing organisms. The organism is supposed to possess at least the following capabilities:[3]

> The ability and, beyond that, the tendency to establish recurrences in the flow of experience; this, in turn, entails at least two capabilities,
> remembering and retrieving (re-presenting) experiences,
> and the ability to make comparisons and judgements of similarity and difference;
> apart from these, there is the presupposition that the organism likes certain experiences better than others, which is to say, it has some elementary values.

The first three of these are indispensable in any theory of learning. Even the parsimonious models of classical and operant conditioning could not do without them. As to the fourth, the assumption of elementary values, it was explicitly embodied in Thorndike's *Law of Effect*: "Other things being equal, connections grow stronger if they issue in satisfying states of affairs" (Thorndike 1931, 1966, p. 101). It remained implicit in psychological learning theories since Thorndike, but the subjectivity of what is 'satisfying' was more or less deliberately obscured by behaviorists through the use of the more objective sounding term 'reinforcement'.

The learning theory that emerges from Piaget's work can be summarized by saying that cognitive change and *learning* take place when a scheme, instead of producing the expected result, leads to perturbation, and perturbation, in turn, leads to accommodation that establishes a new equilibrium. Learning and the knowledge it creates, thus, are explicitly instrumental. But here, again, it is crucial not to be rash and too simplistic in interpreting Piaget. His theory of cognition involves a two-fold instrumentalism. On the *sensory-motor* level, action schemes are instrumental in helping organisms to achieve goals in their interaction with their experiential world. On the level of *reflective abstraction*, however, operative schemes are instrumental in helping organisms achieve a coherent conceptual network that reflects the paths of acting as well as thinking which, at the organisms' present point of experience, have turned out to be viable. The first instrumentality might be called 'utilitarian' (the kind philosophers have traditionally scorned). The second, however, is strictly 'epistemic'. As such, it may be of some philosophical interest – above all because it entails a radical shift in the conception of 'knowledge', a shift that eliminates the paradoxical conception of truth that requires a forever unattainable ontological test. The shift that substitutes viability in the experiential world for correspondence with ontological reality applies to knowledge that results from inductive inferences and generalizations. It does not affect deductive inferences in logic and mathematics. In Piaget's view, the certainty of conclusions in these areas pertains to mental operations and not to sensory-motor material (cf. Beth & Piaget 1961; Glasersfeld 1985b).

THE SOCIAL COMPONENT

In connection with the concept of viability, be it 'utilitarian' or 'epistemic', social interaction plays an important role. Except for animal psychologists, social interaction refers to what goes on among humans and involves language. As a rule it is also treated as essentially different from the interactions human organisms have with other items in their experiential field, because it is more or less tacitly assumed that humans are from the very outset privileged experiential entities. Constructivists have no intention of denying this intuitive human prerogative. But, insofar as their theory of knowing attempts to model the cognitive development that provides the individual organism with *all* the furniture of his or her experiential field, they want to avoid assuming any cognitive structures or categories as innate. Hence, there is the need to hypothesize a model for the conceptual genesis of 'others'.

On the sensory-motor level, the schemes a developing child builds up and manages to keep viable will come to involve a large variety of 'objects'. There will be cups and spoons, building blocks and pencils, rag dolls and teddy bears – all seen, manipulated, and familiar as components of diverse action schemes. But there may also be kittens and perhaps a dog. Though the child may at first approach these items with action schemes that assimilate them to dolls or teddy bears, their unexpected reactions will quickly cause novel kinds of perturbation and inevitable accommodations. The most momentuous of these accommodations can be roughly characterized by saying that the child will come to ascribe to these somewhat unruly entities certain properties that radically differentiate them from the other familiar objects. Among these properties will be the ability to move on their own, the ability to see and to hear, and eventually also the ability to feel pain. The ascription of these properties arises simply because, without them, the child's interactions with kittens and dogs cannot be turned into even moderately reliable schemes.

A very similar development may lead to the child's construction of schemes that involve still more complex items in her experiential environment, namely the human individuals who, to a much greater extent than other recurrent items of experience, make interaction unavoidable. (As we all remember, in many of these inescapable interactions, the schemes that are developed aim at avoiding unpleasant consequences rather than creating rewarding results.) Here again, in order to develop relatively reliable schemes, the child must impute certain capabilities to the objects of interaction. But now these ascriptions comprise not only perceptual but also cognitive capabilities, and soon these formidable 'others' will be seen as intending, making plans, and being both very and not at all predictable in some respects. Indeed, out of the manifold of these frequent but nevertheless special interactions, there eventually emerges the way the developing human individual will think both of 'others' and of him- or herself.

This reciprocity is, I believe, precisely what Kant had in mind when he wrote:

It is manifest that, if one wants to imagine a thinking being, one would have to put oneself in its place and to impute one's own subject to the object one intended to consider.... (Kant 1781, p. 354)

My brief account of the conceptual construction of 'others' is no doubt a crude and preliminary analysis but it at least opens a possibility of approaching the problem without the vacuous assumption of innateness. Besides, the Kantian notion that we impute the cognitive capabilities we isolate in ourselves to our conspecifics, leads to an explanation of why it means so much to us to have our experiential reality confirmed by others. The use of a scheme always involves the expectation of a more or less specific result. On the level of reflective abstraction, the expectation can be turned into a prediction. If we impute planning and foresight to others, this means that we also impute to them some of the schemes that have worked well for ourselves. Then, if a particular prediction we have made concerning an action or reaction of an other turns out to be corroborated by what the other does, this adds a second level of viability to our scheme; and this second level of viability strengthens the experiential reality we have constructed (cf. Glasersfeld 1985a, 1986).

A PERSPECTIVE ON COMMUNICATION

Although it is not always explicitly acknowledged, the separation of two kinds of instrumentality, which I mentioned above, is not a new one in the field of education. Since the days of Socrates, teachers have known that it is one thing to bring students to acquire certain ways of acting – be it kicking a football, performing a multiplication algorithm, or the reciting of verbal expressions – but quite another to engender understanding. The one enterprise could be called 'training', the other 'teaching', but educators, who are often better at the first than at the second, do not always want to maintain the distinction. Consequently, the methods for attaining the two goals tend to be confused. In both, communication plays a considerable part, but what is intended by 'communication' is not quite the same.

Early studies of communication developed a diagrammatic representation of the process as it appears to an outside observer. Success or failure of a communication event was determined on the basis of the observable behavior of a sender and a receiver. This schema was highly successful in the work of communication engineers (Cherry 1966, p. 171). It was also immediately applicable to the behaviorist approach to teaching and learning. The teacher's task, according to that view, consisted largely in providing a set of stimuli and reinforcements apt to condition the student to 'emit' behavioral responses considered appropriate by the teacher. Wherever the goal is students' reliable replication of an observable behavior, this method works well. And, because there is no place in the behaviorist approach for what we would like to call *understanding*, it is not surprising that the behaviorist training rarely, if ever, produces it.

The technical model of communication (Shannon 1948), however, established one feature of the process that remains important no matter from what orientation one approaches it: The physical signals that travel from one communicator to another – for instance the sounds of speech and the visual patterns of print or writing in linguistic communication – do not actually *carry* or *contain* what we think of as 'meaning'. Instead, they should be considered instructions to select particular meanings from a list which, together with the list of agreed signals, constitutes the 'code' of the particular communication system. From this it follows that, if the two lists and the conventional associations that link the items in them are not available to a receiver before the linguistic interaction takes place, the signals will be meaningless for that receiver.

From the constructivist point of view, this feature of communication is of particular interest because it clearly brings out the fact that language users must individually *construct* the meaning of words, phrases, sentences, and texts. Needless to say, this semantic construction does not always have to start from scratch. Once a certain amount of vocabulary and combinatorial rules ('syntax') have been built up in interaction with speakers of the particular language, these patterns can be used to lead a learner to form novel combinations and, thus, novel conceptual compounds. But the basic elements out of which an individual's conceptual structures are composed and the relations by means of which they are held together cannot be transferred from one language user to another, let alone from a proficient speaker to an infant. These building blocks must be abstracted from individual experience; and their interpersonal fit, which makes possible what we call communication, can arise only in the course of protracted interaction, through mutual orientation and adaptation (cf. Maturana 1980).

Though it is often said that normal children acquire their language without noticeable effort, a closer examination shows that the process involved is not as simple as it seems. If, for instance, you want your infant to learn the word 'cup', you will go through a routine that parents have used through the ages. You will point to, and then probably pick up and move, an object that satisfies your definition of 'cup', and at the same time you will repeatedly utter the word. It is likely that mothers and fathers do this intuitively, i.e., without a well-formulated theoretical basis. They do it because it usually works. But the fact that it works does not mean that it has to be a simple matter. There are at least three essential steps the child has to make.

The first consists in focusing attention on some specific sensory signals in the manifold of signals which, at every moment, are available within the child's sensory system; the parent's pointing provides a merely approximate and usually quite ambiguous direction for this act.

The second step consists in isolating and coordinating a group of these sensory signals to form a more or less discrete visual item or 'thing'. The parent's moving the cup greatly aids this process because it accentuates the relevant figure as opposed to the parts of the visual field that are to form the irrelevant ground.[4]

The third step, then, is to associate the isolated visual pattern with the auditory experience produced by the parent's utterances of the word 'cup'. Again, the child must first isolate the sensory signals that constitute this auditory experience from the background (the manifold auditory signals that are available at the moment); and the parent's repetition of the word obviously enhances the process of isolating the auditory pattern as well as its association with the moving visual pattern.

If this sequence of steps provides an adequate analysis of the initial acquisition of the meaning of the word 'cup', it is clear that the child's meaning of that word is made up exclusively of elements which the child abstracts from its own experience. Indeed, anyone who has more or less methodically watched children acquire the use of new words, will have noticed that what they isolate as meanings from their experiences in conjunction with words is often only partially compatible with the meanings the adult speakers of the language take for granted. Thus the child's initial concept of cup often includes the activity of drinking, and sometimes even what is being drunk, e.g., milk. Indeed, it may take quite some time before the continual linguistic and social interaction with other speakers of the language provides occasions for the accommodations that are necessary for the concept the child associates with the word 'cup' to become adapted to the adults' extended use of the word, for instance, in the context of golf greens or championships of the sporting kind.

The process of accommodating and tuning the meaning of words and linguistic expressions actually continues for each of us throughout our lives. No matter how long we have spoken a language, there will still be the occasions when we realize that, up to that point, we have been using a word in a way that now turns out to be idiosyncratic in some particular respect.

Once we come to see this essential and inescapable subjectivity of linguistic meaning, we can no longer maintain the preconceived notion that words *convey* ideas or knowledge; nor can we believe that a listener who apparently 'understands' what we say must necessarily have conceptual structures that are identical with ours. Instead, we come to realize that 'understanding' is a matter of fit rather than match. Put in the simplest way, to understand what someone has said or written means no less but also no more than to have built up a conceptual structure that, in the given context, appears to be *compatible* with the structure the speaker had in mind – and this compatibility, as a rule, manifests itself in no other way than that the receiver says and does nothing that contravenes the speaker's expectations.

Among proficient speakers of a language, the individual's conceptual idiosyncracies rarely surface when the topics of conversation are everyday objects and events. To be considered proficient in a given language requires two things among others: to have available a large enough vocabulary, and to have constructed and sufficiently accommodated and adapted the meanings associated with the words of that vocabulary so that no conceptual discrepancies become apparent in ordinary linguistic interactions. When conversation turns to predominantly abstract matters, it usually does not take long before conceptual discrepancies become noticeable – even among proficient speakers. The discrepancies generate perturbations in the interactors, and at that point the difficulties become insurmountable if the participants believe that their meanings of the words they have used are *true representations* of fixed entities in an objective world apart from any speaker. If, instead, the participants take a constructivist view and assume that a language user's meanings cannot be anything but subjective constructs derived from the speaker's individual experiences, some accommodation and adaptation is usually possible.

From this perspective, the use of language in teaching is far more complicated than it is mostly presumed to be. It cannot be a means of *transferring* information or knowledge to the student. As Rorty says: "The activity of uttering sentences is one of the things people do in order to cope with their environment" (1982, p. XVII). In the teacher's case, language becomes a means of constraining and thus orienting the student's conceptual construction.

This inherent and inescapable indeterminacy of linguistic communication is something the best teachers have always known. Independently of any epistemological orientation, they were intuitively aware of the fact that 'telling' is not enough, because understanding is not a matter of passively receiving but of actively building up. Yet many who are involved in educational activities continue to act as though it were reasonable to believe that the verbal reiteration of facts and principles must eventually generate the desired understanding on the part of students.

CONSEQUENCES FOR EDUCATION

The contemporary movements in the philosophy of science converge in the realization that knowledge must not be considered an objective representation of an external observer-independent environment or world. To paraphrase Rorty, the fact that scientific knowledge enables us to cope does not justify the belief that scientific knowledge provides a picture of the world that corresponds to an absolute reality. This stance tends to suggest a return to the sceptics' age-old assertion that we cannot attain certain knowledge about the world. Educators are traditionally averse to accepting such a view, and it is in this regard that pragmatism and constructivism may play a helpful role.

Both these orientations aim at overcoming the sceptics' pessimism, not by contradicting the assertion that objective knowledge is impossible, but by changing the concept of knowledge. Instead of presupposing that knowledge has to be a 'representation' of what exists, they posit knowledge as a mapping of what, in the light of human experience, turns out to be feasible. If the theory of knowing that constructivism builds up on this basis were adopted as a working hypothesis, it could bring about some rather profound changes in the general practice of education.

First of all, the distinction of utilitarian and epistemic instrumentality would sharpen the distinction between training and learning. It would help to separate the acquisition of skills, i.e., patterns of action, from the active construction of viable conceptual networks, i.e., understanding. Hence it would encourage educators to clarify the particular goals they want to attain. Curricula could be designed with more internal coherence and, consequently, would be more effective, once they deliberately separated the task of achieving a certain level of performance in a skill from that of generating conceptual understanding within a given problem area. There is no question that the old stand-bys 'rote learning' and 'repeated practice' have their value in training, but it is naive to expect that they must also generate understanding.

The analysis of the process of linguistic communication shows that knowledge cannot simply be transferred by means of words. Verbally explaining a problem does not lead to understanding, unless the concepts the listener has associated with the linguistic components of the explanation are compatible with those the explainer has in mind. Hence it is essential that the teacher have an adequate model of the conceptual network within which the student assimilates what he or she is being told. Without such a model as basis, teaching is likely to remain a hit-or-miss affair.

From the constructivist perspective, 'learning' is the product of self-organization. Piaget's dictum "intelligence organizes the world by organizing itself" (1937, p. 311) was a challenge to direct the attention of psychologists to the question of how the rational mind organizes experience and to design a model of this process. His scheme theory, as I outlined it above, is an attempt to answer part of that question. It can be summarized in the statement: Knowledge is never acquired passively, because novelty cannot be handled except through assimilation to a cognitive structure the experiencing subject already has. Indeed, the subject does not perceive an experience as novel until it generates a perturbation relative to some expected result. Only at that point the experience may lead to an accommodation and thus to a novel conceptual structure that re-establishes a relative equilibrium. In this context, it is necessary to emphasize that the most frequent source of perturbations for the developing cognitive subject is the interaction with others.[5] This, indeed, is the reason why constructivist teachers of science and mathematics have been promoting 'group learning', a practice that lets two or three students discuss approaches to a given problem, with little or no interference from the teacher.

Insofar as learning and knowledge are instrumental in establishing and maintaining the cognizing subject's equilibrium, they are *adaptive*. Adaptedness, from the constructivist point of view, must be understood as the condition of fit or viability within external and internal constraints. Constraints, however, effect a negative selection. They block and thus determine what does *not* fit. They do not prescribe the character of what does not collide with them and therefore slips through. Once this way of thinking takes root, it changes the teacher's view of 'problems' and their solution. No longer would it be possible to cling to the notion that a given task has one solution and only one way of arriving at it. The teacher would come to realize that what he or she presents as a 'problem' may be seen differently by the student. Consequently, the student may produce a sensible solution that makes no sense to the teacher. To be then told that it is *wrong* is unhelpful and inhibiting (even if the 'right' way is explained), because it disregards the effort the student put in. Indeed, such bleak corrections are bound to diminish the student's motivation in future attempts. In contrast, constructivist teachers would tend to explore how students see the problem and why their path towards a solution seemed promising to *them*. This in turn makes it possible to build up a hypothetical model of the student's conceptual network and to adapt instructional activity so that it provides occasions for accommodations that are actually within the student's reach.

Fleck's statement that I quoted at the beginning, to the effect that the choice of problems is subject to the 'style' of the scientific community, applies no less to the individual. The character and structure of what an individual *sees* as a problem is under all circumstances determined by the conceptual network and the goals of that individual. Once we adopt this as the working hypothesis, the question of motivation becomes accessible from a new direction. We may not have to do this as long as the subject matter we want to teach provides obvious advantages on the level of *utilitarian* instrumentality (although even there, it should be clear that what a teacher considers useful will not necessarily be considered useful by students). In the case of topics that pertain to *epistemic* instrumentality, the task of fostering motivation is obviously far more difficult. We shall have to make the students perceive the advantage of mastering conceptual models that have a wider range of applicability and success in their experiential world than the ones they have at the moment. More important still, we shall have to create at least some circumstances where the students have the possibility of experiencing the pleasure of finding that a conceptual model they have constructed is, in fact, an adequate and satisfying model in a new situation. Only the experience of such successes and the pleasure they provide can motivate a learner *intellectually* for the task of constructing further conceptual models.

It boils down to what Ceccato, the Italian pioneer of conceptual analysis, said in a talk about education years ago:

The important thing is to show the child (and nothing changes if we substitute 'the student') the direction in which to go, to teach him to find his own path, to retrace it, and to continue it. Only in this way will he be able to assume a scientific attitude with which he can approach also the things of the mind. (1974, p. 137)

This constitutes a drastic modification of the usual procedure. Yet, where it has been tried, its results are startlingly successful.[6]

Recent developments in the philosophy of science have provided a more adequate way of thinking about how scientists proceed to devise better ways of 'coping' with the world of our experience; it should not be surprising that this analysis is applicable also to the process of education. Students may not have the same particular goals that scientists try to attain. But, unless we assume that they share, with the inventors and developers of the conceptual models we call science, the goal of constructing a relatively reliable and coherent model of their individual experiential worlds, we cannot lead them to expand their understanding. Memorizing facts and training in rote procedures cannot achieve this.

Good teachers, as I said before, have practised much of what is suggested here, without the benefit of an explicit theory of knowing. Their approach was intuitive and successful, and this exposition will not present anything to change their ways. But by supplying a theoretical foundation that seems compatible with what has worked in the past, constructivism may provide the thousands of less intuitive educators an accessible way to improve their methods of instruction.

NOTES

[*] I am indebted to Jack Lochhead and John Clement for their helpful critical comments on the draft of this paper. The work that led to it was supported in part by NSF Grants to IBR, University of Georgia, and SRRI, University of Massachusetts.

[1] I am using 'objective' in this traditional philosophical sense and would not want it confused with the Humpty Dumpty-like definitions Siegel suggested in his 1982 article. Although he introduces a dichotomy, he does not separate the two most common uses of the word: (a) referring to knowledge that purports to describe the world as it is; and (b) knowledge that purports to be intersubjective.

[2] This notion of assimilation seems to be compatible with the view of philosophers of science who maintain that all observation is necessarily 'theory-laden'.

[3] Piaget nowhere lists these presuppositions, but they are implicit in his analysis of conceptual development (cf., for instances, Piaget 1937 and 1967b). Another implication of his theory is that none of these presupposed capabilities necessarily requires the subject's conscious awareness (see my 1982).

[4] Note that, even if the child has co-ordinated sensory signals to form such a 'thing' in the past, each new *recognition* involves isolating it in the current experiential field.
[5] Piaget was often criticized for not taking into account the social interaction of the child. This, I believe, sprang from the fact that his readers tacitly assumed that the social context in which a child develops affects the child in a way that must be essentially different from the physical one. Instead, when Piaget speaks of adaptation, it never excludes adaptation to others. But although he explicitly acknowledged social and especially linguistic interaction here and there in his writings (e.g., 1967b, p. 41), he was, as a rule, less interested in the source of perturbations than in the mechanisms for neutralizing them.
[6] Teaching methods that are explicitly constructivist have been documented for instance in Clement 1987; Cobb et al. 1987; Confrey 1984; Duckworth 1987; Lochhead 1983; Steffe 1986; Steffe et al. 1987; Treffers 1987.

REFERENCES

Anonymous: 1711, 'Osservazioni', *Giornale de' Letterati d'Italia*, Venice 5, article VI.
Beth, E. W. and J. Piaget: 1961, *Epistémologie Mathématique et Psychologie*, Presses Universitaires de France, Paris.
Ceccato, S.: 1974, 'In the Garden of Choices', in C. D. Smock and E. v. Glasersfeld (eds.), *Epistemology and Eduation*, Follow Through Publications, Report 14, Athens, Georgia, pp. 123–40, Italian original presented at *Convegno sull'educazione*, San Marino, Italy, 1973.
Cherry, C.: 1966, *On Human Communication*, 2nd ed., MIT Press, Cambridge, Massachusetts.
Clement, J.: 1987, 'Overcoming Students' Misconceptions in Physics: The Role of Anchoring Intuitions and Analogical Validity', in J. Novak (ed.), *Proceedings of the 2nd International Seminar on Educational Strategies in Science and Mathematics*, Vol. 3, Cornell University, Ithaca, New York, pp. 84–97.
Cobb, P., T. Wood and E. Yackel: 1989, 'Philosophy of Science as a Source of Analogies for Mathematics Education', unpublished paper.
Confrey, J.: 1984, 'An Examination of the Conceptions of Mathematics of Young Women in High School', paper presented at Annual Meeting of the *American Educational Research Association*, New Orleans, 1984.
Duckworth, E.: 1987, *The Having of Wonderful Ideas*, Teachers College Press, Columbia University, New York.
Fleck, L.: 1929, 'Zur Krise der "Wirklichkeit"', *Die Naturwissenschaften*, 17, 23, 425–30. (The excerpts from this paper were translated by E. v. Glasersfeld.)
Fleck, L.: 1935, *Entstehung und Entwiklung einer wissenschaftlichen Tatsache*, Benno Schwabe, Basel, Switzerland; reprinted; Suhrkamp 1980, Nördlingen, Germany.
Foerster, H. von: 1981, *Observing Systems*, Intersystems Publications, Salinas, California.
Glasersfeld, E. von: 1982, 'An Interpretation of Piaget's Constructivism', *Revue Internationale de Philosophie* 36, 612–35.
Glasersfeld, E. von: 1985a, 'Reconstructing the Concept of Knowledge', *Archives de Psychologie* 53, 91–101.

Glasersfeld, E. von: 1985b, 'Representation and Deducation', in L. Streefland (ed.), *Proceedings of the 9th Conference for the Psychology of Mathematics Education*, vol. 1, State University of Utrecht, The Netherlands, pp. 484–89.

Glasersfeld, E. von: 1986, 'Steps in the Construction of "Others" and "Reality"', in R. Trappl (ed.), *Power, Autonomy, Utopia*, Plenum, London, pp. 107–16.

Hebb, D. O.: 1958, 'Alice in Wonderland or Psychology among the Biological Sciences', in Harlow and Woolsey (eds.), *Biological and Biochemical Bases of Behavior*, University of Wisconsin Press, Madison, pp. 451–67.

Kant, I.: 1781, *Kritik der reinen Vernunft, Gesammelte Schriften, Band IV*, Königlich. Preussische Akademie, Berlin, pp. 1910ff.

Kuhn, T. S.: 1970, *The Structure of Scientific Revolutions*, 2nd ed., University of Chicago Press, Chicago, first published 1962.

Lochhead, J.: 1983, 'Constructivist Approaches to Teaching Mathematics and Science at the College Level', in J. Bergeron and N. Herscovics (eds.), *Proceedings of 5th Annual Meeting of the North American Group for the Psychology of Mathematics Education*, Vol. 2. PME-NA, Montreal, Canada, pp. 74–80.

Maturana, H. R.: 1980, 'Biology of Cognition', in H. R. Maturana and F. J. Varela (eds.), *Autopoiesis and Cognition*, D. Reidel, Dordrecht, pp. 5–58.

Nash, J.: 1970, *Developmental Psychology*, Prentice-Hall, Englewood Cliffs, New Jersey.

Piaget, J.: 1937, *La construction du réel chez l'enfant*, Delachaux et Niestlé, Neuchâtel, Switzerland.

Piaget, J.: 1967a, *Biologie et connaissance*, Gallimard, Paris.

Piaget, J.: 1967b, *Six Psychological Studies*, Vintage Books, New York.

Piaget, J. and R. Garcia: 1983, *Psychogénèse et histoire des sciences*, Flammarion, Paris.

Pittendrigh, C. S.: 1958, 'Adaptation, Natural Selection, and Behavior', in A. Roe and G. G. Simpson (eds.), *Behavior and Evolution*, Yale University Press, New Haven, pp. 390–416.

Shannon, C.: 1948, 'The Mathematical Theory of Communication', *Bell Systems Technical Journal* **27**, 379–423, 623–56.

Siegel, H.: 1982, 'On the Parallel between Piagetian Cognitive Development and the History of Science', *Philosophy of Social Science* **12**, 375–86.

Steffe, L. P.: 1986, 'Principles of Mathematical Curriculum Design in Early Childhood Education', Symposium on Early Mathematics Learning: Teacher-focused Curriculum Change, *American Educational Research Association*, Washington, DC.

Steffe, L. P. and P. Cobb: 1987, *Construction of Arithmetical Meanings and Strategies*, Springer-Verlag, New York.

Thorndike, E.: 1966, *Human Learning*, MIT Press, Cambridge, Massachusetts, first published, 1931.

Treffers, A.: 1987, *Three Dimensions: A Model of Goal and Theory Description in Mathematics Education – The Wiskobas Project*, D. Reidel, Dordrecht.

Vico, G.: 1710, *De antiquissima Italorum sapientia*, with Italian translation by F. S. Pomodoro, Stamperia de' Classici Latini, Naples, 1858.

Scientific Reasoning Research Institute
University of Massachusetts
Amherst, MA 01003
U.S.A.

Constructivism in Science and Science Education: A Philosophical Critique

ROBERT NOLA

Department of Philosophy, University of Auckland, Private Bag 92019, Auckland, New Zealand

> Two or three times in this author's arguments I have noticed that in order to prove that matters stand in such-and-such a way, he makes use of the remark that in just this way do they accommodate themselves to our comprehension, and that otherwise we should have no knowledge of this or that detail; or that the criterion of philosophizing would be ruined; as if nature first made the brain of man, and then arranged everything to conform to the capacity of his intellect. But I should think rather that nature first made things in her own way, and then made human reason skillful enough to be able to understand, but only by hard work, some part of her secrets.
>
> Galileo Galilei, *Dialogue Concerning the Two Chief World Systems*, Stillman Drake trans., University of Calfornia Press, pp. 264–5.

ABSTRACT: This paper argues that constructivist science education works with an unsatisfactory account of knowledge which affects both its account of the nature of science and of science education. The paper begins with a brief survey of realism and anti-realism in science and the varieties of constructivism that can be found. In the second section the important conception of knowledge and teaching that Plato develops in the *Meno* is contrasted with constructivism. The section ends with an account of the contribution that Vico (as understood by constructivists), Kant and Piaget have made to constructivist doctrines. Section three is devoted to a critique of the theory of knowledge and the anti-realism of von Glasersfeld. The final section considers the connection, or lack of it, between the constructivist view of science and knowledge and the teaching of science.

1. VARIETIES OF CONSTRUCTIVISM

'Constructivism: A Paradigm for the Practice of Science Education'[1] is the heading to Tobin's 'Preface' to his collection of papers by educationalists largely committed to the cause of constructivism in science education. He begins by saying that 'currently there is a paradigm war raging in education'[2] in which some have 'argued for a change in epistemology', evidently with some success for he claims 'there is evidence of widespread acceptance of alternatives to objectivism, one of which is constructivism'.[3] But what are constructivism and its alternative objectivism?

The contrast between realism and anti-realism, of which constructivism is a variety, is an old one in the theory of scientific knowledge. Objectivism is a broad doctrine which includes not only scientific realism (a doctrine discussed in this paper), but also the idea that there are objective critical methods for adjudicating between scientific hypotheses (an aspect of objectivism only obliquely mentioned here). Does science make discoveries about a human-independent world, including the world of unobservable entities such as gravitation or electric charge? Scientific realists say 'Yes' while admitting that we are fallible and may not always be right about what exists in the unobservable realm. Common sense realism and scientific realism maintain that there exists objects, events and processes in the world which are independent of all human perception and all thought or theorizing about them. They are independent in the sense that if there were no humans around to perceive or think about them then they would still exist. Such items are not constructs of ours, or projections onto the world by us; nor do we constitute them in some way by our thought or theorizing about them. Are there such items? Common sense realists maintain that the Sun, cats, water, etc exist in a mind-independent way (though we have been mistaken about some items in the past such as witches, goblins, etc.). Scientific realists maintain that science has discovered (not invented or constructed) items such as electrons, viruses, tectonic plates, galaxies, etc (though scientists have been mistaken about some items in the past such as epicycles, electromagnetic ether, phlogiston, polywater, etc.). It is an additional matter to also say that our theories of science are true.[4] Some realists resist this saying either that it is not necessary to claim truth for our theories as well as the commonsense and scientific realism just defined; or they say that at best our theories are idealizations of, or approximations to, what goes on at the level of the observable and the unobservable.[5]

Anti-realists, including constructivists, deny that science makes discoveries about a human-independent world, including the world of unobservable entities – but they qualify this in various ways. Constructivists allege that it is we who constitute or construct, on the basis of our theorizing or our experience, the allegedly unobservable items postulated in our theories. In a different vein Thomas Kuhn expresses scepticism about our ever being able to get at the truth about what is 'really there' saying: 'the notion of a match between the ontology of a theory and its real counterpart in nature now seems to me illusive in principle'. (Kuhn 1970, p. 206) Bas van Fraassen is a leading philosopher of science who embraces what he calls 'constructive empiricism', the view that in science we aim for models which are only required to fit the observable phenomena. Realists, he says, illegitimately aim for more; they want theories which not only fit the observable phenomena but are also true of unobservable reality.[6] Van Fraassen's brand of constructive empiricism involves technicalities; since it has not found favour with constructivists in science education it will not be discussed here. Some anti-realists are merely sceptical of any claims that we can know anything beyond experience; hence an empiricism with respect to the entities postulated in science made famous in various ways by the eighteenth century empiricist philosophers Hume and Berkeley and twentieth century logical positivists.

What has the venerable philosophical debate between realists and anti-realists about scientific knowledge to do with the teaching and learning of science? There is no necessary connection. Constructivists in science education often wrongly assume that the debate can tell us something about the teaching and learning of science. Constructivist teaching and learning is another matter, best contrasted with didacticism. However it is commonly assumed that a realist account of science goes with a didactic 'tell it how it is' approach to teaching and learning while a non-realist account goes with a more personal constructivist approach. But these assumptions are too crude. The main topic of this paper concerns the way in which philosophy bears on the nature of scientific knowledge and on science education, and some of the misleading links constructivists allege hold between the two.

Constructivism in education theory is a protean doctrine in which the metaphors of building and inventing have run riot. Four questions initially strike one. Who does the constructing? What is constructed? What is the relation of constructing? How does one construct? Sometimes it is the individual person who does the constructing, as in more psychologically oriented accounts influenced by the work of Piaget or Von Glasersfeld. But we are not to confuse the constructing in which pupils may engage in learning science with the constructing scientists may engage while actively doing science. It might be pedagogically useful for some pupils to follow, in their learning, the actual path of the evolution of some science; but deep confusion can only result from not separating scientists' alleged 'construction' of scientific knowledge from pupils' 'constructivist' learning, or teachers' 'constructivist' teaching, of science.

Sometimes it is a group (e.g., pupils and teacher) or society as a whole which constructs. There are several sources of social constructivism in science education, from the social theories of language of Vygotsky and Wittgenstein to the social theories of science of the Edinburgh Strong Programme for the Sociology of Scientific Knowledge, advocated in various ways by Bloor, Barnes, Collins, Pinch, Schaffer, Shapin and a host of others. There are also social constructivist elements in Piaget and Von Glasersfeld. However, once more, confusion between pupils' and scientists' 'social construction' must be avoided.

What is constructed? A host of items from models, hypotheses, theories and knowledge in both science and mathematics as well as the entities postulated in science (e.g., electrons, genes, etc.) and mathematics (e.g., numbers, geometrical figures, etc.); broad mentalistic categories such as ideas, experience, meanings, beliefs and languages; and on the pedagogical side pupils' knowledge, teaching science, learning science, curricula, the development of curricula and research programmes for educationalists. We will be concerned here only with the alleged scientists' construction of scientific knowledge out of experience and of the alleged pupils' constructivist learning of scientific knowledge from their experience. Since I maintain that constructivists systematically misunderstand the nature of knowledge much of this paper will be concerned with the central epistemological concept of knowledge.

What sort of relation is constructing, and how does one do it? Most of the difficulties in constructivism can be located here. One influential account of the constructivist educational 'paradigm' has been advocated by Ernst von Glasersfeld; his distinction between *trivial* and *radical* constructivism is explored in section 3.

2. PHILOSOPHICAL ORIGINS: A BRIEF SURVEY

2.1. *Socrates and Plato: the Construction of Reasons for Knowledge*

Down the ages the works of Plato have been an important source for educationalists. The *Meno* provides us not only with the first attempt at a definition of knowledge as opposed to belief but also, in Socrates' encounter with the Slave-Boy, a model for pedagogy which is non-didactic and thus important for both constructivists and non-constructivists alike. Consider the episode with the Slave-Boy.

Meno presents Socrates with a conundrum part of which says that we cannot learn anything new because we can not know when we have hit upon the right answer (*Meno* 80D-E). Socrates counters this in the dialogue with a Slave-Boy who patently does not know the answer to the geometrical question 'What is the length of the side of a square double the area of a given square the side of which is 2 metres long' and who then comes to know what is the correct answer. The Boy initially thinks that the answer is double the side of the given square, i.e., 4 m. Using a question-answer method, Socrates gets the Boy to work out the areas of the two squares whose sides are 2 m and 4 m long. While doing this Socrates emphasises his non-didactic approach towards the Boy's thinking about the geometrical problem when he say to Meno: 'You see Meno that I am not teaching [telling] him anything, only asking' (84E). Socrates does not *tell* the Slave-Boy that his answer of 4 m is wrong; rather through the question-answer method the Boy *comes to realise himself* that his answer is wrong. The same non-didactic procedure is repeated to show the Boy that his next answer of 3 m is also wrong.

After two wrong answers the Boy becomes perplexed; he does not know what other answer might be correct and, moreover, he knows that he does not know. Socrates turns to Meno and says:

> Now notice what, starting from the state of perplexity, he will discover by seeking the truth in company with me, though I simply ask him questions without teaching him. Be ready to catch me if I give him any instruction or explanation instead of simply interrogating him on his own opinions (84C-D).

Since the Boy has run out of suggestions Socrates gives him a hint about what geometrical move he should try next and then, employing only the Boy's capacities to reason about the questions put to him, takes him to the correct conclusion, viz., the diagonal of the given square of side 2 m.

Two important points are now made in the dialogue about the pedagogical process the Boy has undergone. Socrates asks Meno: 'Has he, the Slave Boy, answered with any opinions that were not his own'. The reply is 'No' (85B). The first point is that at each stage of the questioning process the Slave-Boy has acquired, or, if you like the metaphor has 'constructed', for himself the reasons which show that his first two answers were wrong and that his third answer is correct. In being opposed to didactic teaching and learning of reasons which underpin knowledge, Socrates and Plato are the first constructivists in education. This is part of the common sense core of constructivism that few would deny. But, as will be argued, even though Socrates and Plato could admit a 'constructive' element in arriving at the reasons which eliminate false belief and turn true belief into knowledge, they do not endorse a constructivist account of the nature of knowledge itself.

The second point concerns knowledge directly. Socrates comments on the correct answer given by the Slave-Boy:

> At the present these opinions [that the answer is the diagonal] being newly aroused, have a dream like quality. But if the same questions are put to him on many occasions and in different ways, you can see that in the end he will have a knowledge on the subject as accurate as anybody's.... This knowledge will not come from teaching but from questioning. He will recover it for himself (85C-D).

In Socrates' view, students do not acquire knowledge through picking up bits of (true) information didactically conveyed to them. Even being led through a question-answer session does not provide, by itself, knowledge; at best the process can only leads pupils to the correct belief. Only when they can go through the steps of reasoning by themselves and thereby make fully explicit to themselves the reasons for the correct answer will they have knowledge. Re-expressing this more metaphorically, only by 'constructing' for oneself the reasons for a true belief can one acquire knowledge. (Of course this does not show that no knowledge whatever can be acquired through telling someone something or conveying information to them; rather, in this case and many others in science and mathematics, discovering the reasons for something is an important necessary condition for acquiring knowledge.)

Socrates' answer to the initial conundrum posed by Meno is that we can recognise that we have knowledge when we have satisfactory reasons for the truth of what we believe. This is spelled out by Socrates as follows:

> True opinions are a fine thing and do all sorts of good so long as they stay in their place; but they will not stay long. They run away from a man's mind, so they are not worth much until you tether them by working out the reason.... Once they are tied down, they become knowledge, and are stable. That is why knowledge is something more valuable than right opinion. What distinguishes one from the other is the tether (97E–98A).

The definition of knowledge proposed here is important. Expressed explicitly, where A is some person and 'p' stands for the content of a belief held by A:

A knows that p = Defn
1. p is true (Truth Condition);
2. A believes that p (Belief Condition);
3. A has a tethering reason (justification, evidence) that p (Justification Condition).

Note that the term 'reason' does double duty in the definition; it can refer to the evidence or ground for p or it can refer to the path of reasoning along which a knower must travel from evidence to p. Reasons and reasoning play a dual role in distinguishing true from false belief and converting true beliefs into knowledge.

Reasons as evidence and reasoning about evidence play a central role in science, the most consciously rational means we have devised for investigating the world. In science reasons and reasoning can be a source of new hypotheses; they can provide the means for testing hypotheses; they tell us why one belief rather than another ought to be held; they help demarcate science from those areas of non-science where reason plays no role in belief. Granted this, reasons should play a central role in science education.[7] But not any reasons will do for knowledge. Some versions of constructivism carry the implication that any kind of construction by a pupil can be permitted. However the anti-didacticism of Socrates is accompanied by *objective* constraints of reasons and reasoning upon knowledge that anyone who hopes to know some science (or anything else) can and must come to grasp.

Surprisingly Plato's tether of reasons goes almost unmentioned in constructivist accounts of knowledge; merely constructing beliefs is often regarded as sufficient for knowledge. By omitting the third condition for knowledge constructivists conflate the definition of knowledge (such as that given above) with the process whereby a person comes to know. As the episode with the Boy illustrates, the process of eliminating error and arriving at knowledge by employing reasons has constructivist elements. In contrast, the definition of knowledge is not a process of constructing; rather it concerns the final product, a state of knowledge, which results from a properly constructed reasoning process.

Socrates insists that one of the few things he knows is that knowledge and true opinion are different (98B). However the account of knowledge given in the *Meno* is also bound up with a theory of knowledge as the recollection of what has been implanted in our immortal souls. Even though he expounds this doctrine Socrates says 'I shouldn't like to take my oath on the whole story' (86B) – and most commentators would argue that he is right to be cautious. However the recollection story can be reconstrued as a first primitive account of either innate belief or of *a priori* knowledge. The Slave-Boy, like all pupils, must have at least some innate capacity to reason about geometrical problems since he has been taught neither how to reason nor how to recognise Plato's tethering reasons as reasons for belief. As a teacher, Socrates starts with the abilities and knowledge the pupil already posses, something constructivists and non-constructivists alike applaud. However the ability to reason and to recognise reasons for and against our beliefs is also something which can be taught.

Plato's tether definition of knowledge will be adopted in what follows. However this was not Plato's final word on the matter; a critical development of his earlier *Meno* definition can be found in his later *Theaetatus*.[8] Nor is the definition endorsed by modern epistemologists who find inadequate the simple reason or justification condition expressed in clause (3) of the definition.[9] However none of the modifications that epistemologists make to the definition will be of comfort to constructivists since they underline even more strongly inadequacies in their conception of knowledge which ignores the role of evidence and reasons. Finally the definition does not cater for knowledge which does not require reasons; this is particularly the case for knowledge based directly on perceptual experience. So let us add what the definition omits, viz., that there is knowledge without reasons based on direct experience. Such an admission should cause no difficulty in the dialectic between constructivists and non-constructivists since it is common ground; constructivists also speak of knowledge as based on 'constructions' out of experience.

2.2. *Vico and Knowledge as Construction in Science Education*

Strictly this section is more about Vico as he is represented and used in constructivist theories of science education than about Vico himself. A leading constructivist, Ernst von Glasersfeld, cites the eighteenth century Italian philosopher Giambattista Vico as a significant precursor of constructivism when in a brief summary he tells us:

> One of Vico's basic ideas was that epistemic agents can *know* nothing but the cognitive structures they themselves have put together. He expressed this in many ways, and the most striking is perhaps: '*God is the artificer of nature, man the god of artefacts*'. Over and over he stresses that 'to know' means *to know how to make*. . . . one knows a thing only when one can tell what components it consists of. Consequently, God alone can know the *real* world, because He knows how and of what He has created it. In contrast, the human knower can only know what the human knower has constructed. (Von Glasersfeld 1989, p. 123)

But this does not contain a satisfactory definition of knowledge because it is circular; the term to be defined, 'know', is used again in the definition 'to know how to make, or to create' (or 'to tell' which is an epistemic notion).

Vico's central thesis states: A knows x if and only if A knows how to make (create) x. But this is inadequate because it rides roughshod over several important distinct grammatical constructions concerning the word 'know' commonly made in the literature on epistemology. Isaiah Berlin in an essay on Vico[10] attempts to come to terms with Vico's formula that 'knowing is a knowing how to make', settling on the idea that the knowledge involved is a form of actors' knowledge based on direct acquaintance with the situation in which they find themselves. Actors in their situations have 'insiders' knowledge of their thoughts, desires, beliefs and situation as opposed to the 'outsiders' perspective which is at best knowledge by inference of these things. Berlin's point is useful; but knowledge by direct acquaintance is only one of several varieties of knowledge that any full account of knowledge must set out. The following lists six kinds of knowledge on the basis of the grammatical constructions into which the word 'know' enters.

1. The word 'know' can take a direct object as when the blank in 'a person knows . . .' is filled by a noun expression; this is commonly called 'knowledge by acquaintance'. For example, a person may be said to know the way home, or to know an apple (when one sees one) or to know the Pope, or to know their own feelings, thoughts, etc. In these cases a person knows an object O in the sense of being able to identify, or pick out, O or to know O by inspecting the contents of their own conscious experience such as their feelings, desires, beliefs, etc.
2. A person 'knows *how to* . . .' where the blank is filled by an expression referring to a skill or ability. For example, the person knows how to play a violin, how to speak Russian, how to differentiate an equation, etc.
3. A person 'knows *how* . . .' as when a person knows how a diesel engine works, how the blood carries oxygen, etc. In these cases a person has knowledge in the sense that they can offer an explanation. Explanation also underlies the next kind of grammatical construction of 'know'.
4. A person can know *why*, say, the Earth is an oblate sphere, or *why* aspirin cures headaches, etc.
5. Cognitive knowledge is commonly expressed in the form 'a person knows *that* . . .', the blank being filled by a sentence; e.g., 'the Earth rotates on its axis', 'aspirin cures headaches', etc.
6. Finally, knowledge of definitions or natures is expressed in the form 'a person knows *what* . . . is', where the blank can be filled by the name of a sort of thing, e.g., 'a gene', oxygen', 'a triangle', '"π" stands for in mathematics', etc.

Clearly all these uses of 'know' cannot be reduced to a version of (1) as in 'know how to create', or a version of (2), understood as actor's knowledge by direct acquaintance. Moreover such conflation ruins the fine distinctions we make in our ordinary use of 'know'. A person can make all of the following 'knowledge' claims: I know aspirin (i.e., can identify it by, say, label); I know *how* to take aspirin (by swallowing); I do not know *why* aspirin relieves headaches; I do not know *how* aspirin works; I do know *that* aspirin cures headaches; I do not know *what* aspirin is. Clearly constructivists who follow Vico would do serious damage to the concept of knowledge by ignoring such a range of epistemic notions. This has important consequences for education; for each of the constructions above in which the word 'know' appears there will be a corresponding different kind of teaching and learning involved, e.g., teaching and learning how..., that..., what..., etc.

By ignoring the variety of uses of 'know', counter-examples to Vico's thesis can be readily generated. For a long time humans knew (could identify) honey but only bees knew how to make honey. Again, as astronomy advances we get to know a lot about the origin of the solar system or the big bang. But even if we were to have full knowledge of these, we will hardly be in a position to make either a solar system or a big bang. Conversely, Fleming knew how to make penicillin (by producing more of the original culture in other dishes) without 'knowing' penicillin, i.e., knowing what it is (this was investigated later by Florey and Chain). Other difficulties in Vico's conception of knowledge are thoroughly and critically examined in Suchting (1992, pp. 232–39).

Constructivists use Vico's thesis to make two related claims: only God can know the natural world since He made it; since we did not make the world we cannot know it. The second claim has important consequences for the constructivist view of science. It must follow that since we did not make the world investigated in physics, chemistry, biology, etc., we cannot know those aspects of the world that these sciences purportedly investigate. This casts a sceptical veil over many of the important discoveries about the unobservable world that have been made in science from protons and electric fields to genes, viruses and pulsars. As we will see constructivists are willing to embrace this consequence thus endorsing a thoroughgoing non-realist understanding of science.

What do constructivist followers of Vico think we humans know? Only what we can construct or create out of... what? For constructivists what we build are 'cognitive structures'; these, we may assume, include beliefs, particularly scientific beliefs which may be observations, hypotheses or theories. The building blocks can only be items such as meanings, ideas, and most importantly for science, experiences. Depending on how strictly one takes the building-block idea, those who put emphasis on construction of beliefs out of experience join the company of empiricists such as Locke, Berkeley and Hume and twentieth-century logical positivists and phenomenalists. Eighteenth century philosophers embarked on a programme, initiated by Descartes, of showing how, from perceptual experiences, it might be possible to give an account of the external world. Notoriously Berkeley alleged that no account was possible; his position, which denies the realist's independently existing world, is variously known as instrumentalism when applied to scientific knowledge, or as idealism when applied to our knowledge generally. Twentieth century logical positivists tried to show how the external world and the unobservable postulates of science are *logical constructs* out of experience (hence the label 'logical positivist'). Talk of 'logical construction' gives a fairly precise sense to the method of construction employed by empiricists. Though constructivists in science education are not logical constructivists, it is incumbent upon them to elaborate the methods of construction to be employed in their epistemology – something that is not easy to find in their work. What will be illustrated in section 3 is that constructivism is a close ally of empiricism and, in the case of Von Glasersfeld, Berkeley's anti-realist idealism.

The quotation from Vico expresses a thesis which we can dub 'Vico's central Epistemic Principle' (VEP): we epistemic agents can know only those thoughts, beliefs, ideas or 'cognitive structures' that we have put together. On the face of it this merely expresses one of the necessary conditions for knowledge: if we know something then we have formed some beliefs (cognitive structures) for ourselves. On Plato's tether definition of knowledge this is trivially correct. But often this is wrongly taken to also give a *sufficient* condition for knowledge. Understood this way VEP is not something we should accept. The history of human thought is replete with ingenious 'cognitive structures' such as those put together in myth, religion and current urban myths such as UFO visitations. Even in the history of science we can find many beliefs that do not count as knowledge, e.g., beliefs about Ptolemy's epicyclic theory of the planets, the phlogiston theory of combustion, the ether theory of light transmission, the existence of polywater, the efficacy of cold fussion, and so on. And epistemologists and psychologists have been at pains to get us to exclude from the realm of knowledge ideas which arise from perceptual illusions and dreams.

Constructivists who adopt VEP as sufficient for knowledge say that a pupil has knowledge as soon as the pupil puts some 'cognitive structures' together. But at best this can apply only to some beliefs about current experience or memories of previous experiences. However most beliefs are about matters that transcend current or past experience, such as the beliefs found in science that the Earth rotates on its axis, or that dinosaurs became extinct millions of years ago, or that like charges repel like charges, etc. Plato and most epistemologists would say that, for knowledge of other than immediate experience, while it is necessary to put some 'cognitive structures' together this is definitely not sufficient. What is missing is mention of any reasons or justification for such 'cognitive structures'. As we will see, it is one of the characteristics of constructivism to accept VEP as a necessary and sufficient account of knowledge thereby ignoring Plato's tether requirement altogether.

2.3. *Kant: Construction, Invention and Test*

Some constructivists take comfort from a particular (mis)reading of Kant's *Critique of Pure Reason*. In the 'Preface to the Second Edition' Kant says of geometers that it is not sufficient for them to obtain knowledge of the properties of, say, an isosceles triangle merely by looking at such a triangle drawn on a page, or by meditating on the concept of such a triangle. For knowledge 'it is necessary to produce these properties, as it were, by a positive *a priori construction*'; and the geometer 'must not attribute to the object any other properties than those which necessarily followed from that which he had himself, in accordance with his conception, placed in the object'. Concerning the empirical sciences, Kant mentions the work of Galileo, Torricelli and Stahl and says: 'reason only perceives that which it produces after its own design; that it must not be content to follow ... in the leading-strings of nature but must proceed in advance with principles of judgement according to unvarying laws and compel nature to answer its questions' (Kant 1934, p. 10).

Consider Kant's comments on geometry first. Constructivists in mathematics influenced by Kant, view mathematical knowledge as based in objects which we construct *a priori*, where '*a priori*' is understood to mean 'independently from any constraints of experience'. What further knowledge we have of them arises only from what we draw from what we construct independently of experience – it cannot be drawn from anything found in experience. Similarly constructivists in the empirical sciences influenced by Kant recommend us not to follow nature slavishly as do naive inductivists whose knowledge is based only in experience and what they can infer from experience. Rather we are to anticipate nature with our own scientific constructions of hypotheses or theories. Such a constructivist scientist, as Kant puts it, is not like a 'pupil who listens to all that the master chooses to tell him', but is like a 'judge who compels the witness to reply to those questions which he himself thinks fit to propose' (*ibid*, pp. 10–11).

Kant's 'constructivist' view of empirical science bears an affinity with one of the classical methodologies of science, viz., the hypothetico-deductive method. In this method hypotheses are initially proposed, conjectured or freely invented; or more metaphorically we can say that they are 'constructed' (though the metaphor can mislead). The conjectures must then be tested by drawing out their consequences which, in turn, act as a guide to appropriate evidence; the evidence when found may say either 'Yes' or 'No' to the consequences. Such a model relies implicitly on a distinction, endorsed by Popper and Reichenbach,[11] between the context of discovery or invention and the context of justification. We freely invent hypotheses in science, i.e., we 'construct' them. According to Popper and Reichenbach the context of discovery/invention is not subject to logical analysis and is part of the psychology or sociology of invention. In contrast once the freely 'constructed' hypotheses are made public then the task of justifying them arises, such justification involving objective methods of test. Tests, according to the hypothetico-deductive method, involve the application of hypotheses in concrete cases and the drawing out of consequences which can then be tested against what is observed. In this Nature is a judge which can say either 'Yes' or 'No' to the consequences drawn.

It is important to consider the case where Nature says 'No'. What should we then say about the constructed hypotheses or theories? Kant and the constructivists say little about this central matter in theories of scientific method. Consider only Popper's reply to this question, which turns on central tenets of his falsificationist philosophy of science. Popper agrees with constructivists in part when he says in the epigram to *The Logic of Scientific Discovery*: 'Hypotheses are nets; only he who casts will catch'. The metaphor of 'constructing' is recaptured in the 'casting of nets'. However Popper distances himself from constructivists when he insists that we must subject any hypotheses we cast to severe critical examination. Importantly he says of Kant:

> When Kant said that our intellect imposes its laws on nature, he was right – except that he did not notice how often our intellect fails in the attempt; the regularities we try to impose are *psychologically a priori*, but there is not the slightest reason to suppose that they are *a priori valid*.[12]

In Popperian, and most other, methodology, if our constructions fail to pass sufficient tests then they are falsified.

For Popper the constructivist metaphor could be employed in two ways – first in the initial construction of the scientific theory to be tested and, second, in the construction of the tests the theory is to undergo. This procedure can have a counterpart in science education; the activity of the scientist in inventing and testing theory may be re-constructed by pupils when they come to learn about the relevant episode in science.

But once again the problem of Plato's tether comes to haunt those constructivists who say too little about what goes wrong either when scientists construct false theories or when pupils construct wrong beliefs for themselves about science. For Popper we tentatively tether scientific hypotheses that pass severe tests but we are to drop the tether when they fail them. We have no reason to accept those constructions to which nature says 'No'. The same must hold for pupil learning. Since students can learn what is false as well as what is true, then correction becomes imperative. How correction is made is another matter; one way has been discussed in the Socratic question-answer method outlined in section 2.1.

2.4. *Piaget: Knowledge as Construction and as Adaptation*

Von Glasersfeld takes his cue concerning knowledge from both Vico and Piaget when he says:

> As for Vico, knowledge for Piaget is never (and can never be) a 'representation' of the real world. Instead it is the collection of conceptual structures that turn out to be adapted or, as I would say, *viable* within the knowing subject's range of experience (Von Glasersfeld 1989, p. 125).

Piaget's fundamental idea, developed over many publications especially his *Biology and Knowledge*, is that knowledge is merely another stage in the adaptation of an organism to its environment. The central tenet of evolutionary biology is that each organism (or species) regulates itself in its environment in order to ensure its survival. For those organisms with cognitive functioning (which depends on a sufficiently developed nervous system) Piaget proposes a guiding hypothesis about the link between adaptation at the organismic and cognitive levels: 'cognitive processes . . .

are the outcome of organic autoregulation, reflecting its essential mechanisms'. In the light of this hypothesis Piaget asserts for the cognitive process of knowing: 'knowledge is not a copy of the environment but a system of real interactions reflecting the autoregulatory organisation of life just as much as the things themselves do' (Piaget 1971, section 3.1). Later Piaget puts this more crisply when he emphasises that neither knowledge nor truth copy reality: 'knowledge is *essentially construction*' (*ibid.*, section 23.1) Hence the link between Piaget and one of the central tenets of constructivism in education.

Piaget's overall position is that of a scientific realist in that he sets his theory of cognitive functioning within the theory of evolution in biology. He understands this theory to tell us something about how the mind-independent world works, something a scientific realist can applaud. However it is Piaget's epistemology which leads him away from claims about knowledge of an independent world (as understood in Plato's tether definition) to a constructivist viewpoint. Philosophers can rightly complain that Piaget has overlooked one of the important cognitive functions of most of the higher animals, viz., belief, and that most of what Piaget says of knowledge in the above passages and elsewhere should more correctly be said of belief. In support of this the reader will find that the index of *Biology and Philosophy* has copious entries under 'Knowledge' but none under 'Belief'. The following may be part of the explanation why Piaget fails to make the crucial distinction between knowledge and belief. Plato argues correctly in the *Meno* (97–98) that there is no difference, as far as the practical outcome of our actions is concerned, between acting on a true belief and acting on knowledge. So, at the biological level it will be sufficient for the survival of an organism with cognitive functioning to have (on the whole) merely true beliefs rather than something more, viz., knowledge.[13] The tether for belief is otiose where only the correctness of the belief (for Piaget 'knowledge') is relevant. But this does not fully explain Piaget's conflation of knowledge with belief; as will be suggested, his guiding hypothesis, along with his misleading characterization of knowledge, also contributes to the conflation.

Concerning the definition of knowledge in section 2.1, Piaget makes no mention of the third tether condition which requires reasons. Nor does he give much consideration to the first condition concerning truth. In section 23.1 of *Biology and Knowledge* he explicitly rejects the naive copy theory of truth but accepts that 'it [truth] is an organization of the real world'. But this is seriously problematic. We need to ask: whose organisation, and organisation of what? On the one hand Piaget wants to avoid, as he puts it, the unacceptably relativistic doctrines of 'anthropo- or sociocentrism' in the claim that the 'organisation' can only be done by humans. But on the other hand he wants his truths to be about the world that science investigates without our organisational activities (which he thinks are biological) in any way hindering access to that world. Piaget fails to resolve his philosophical difficulties about truth and knowledge despite the epistemological rabbit he pulls out of the hat at the end of his discussion of truth, viz., 'knowledge is essentially construction'. Constructivists influenced by Piaget have inherited all of his philosophical difficulties about truth and knowledge.

Organisms are able to autoregulate themselves in a constantly fluctuating environment in order to achieve equilibrium. Organisms with cognitive systems possess even greater means of autoregulation through what Piaget calls 'equilibration', the ability to compensate cognitively for exterior disturbances. A simple illustration of such autoregulation would be our cognitive ability to make judgements about sameness of actual size of an object despite changes in apparent size and distance as we, or the object, move about. Central to Piaget's theories of equilibration and learning are his notions of *action schemata, assimilation* and *accommodation*.

In line with Vico, Piaget tells us: 'Knowledge does not make a copy of reality but, rather, react[s] to it and transform[s] it . . .' (*ibid.*, section 1.2) For Piaget the processes of learning and knowing involve the active construction of action schemata. To use an illustration of Von Glasersfeld (1989, p. 127), an infant constructs an action schema concerning rattles of the following sort. (1) the infant recognises rattles as something with a particular shape which it can grasp; (2) it associates them with certain activities such as shaking; (3) it has expectations of them when shaken, e.g., noise. Objects may be assimilated to this scheme, successfully in the case of other rattles but unsuccessfully in the case of most other objects, e.g., spoons which fit (1) and (2) but not (3) of the schema. The lack of assimilation of the spoon to the action schema above leads to the recognition of the spoon as a non-rattle and perhaps its accommodation to a new assimilation action schema (of, say, banging spoons on tables, something which cannot be done with a stuffed Teddy-bear, and so on). In sum, the equilibrium with the outside world is upset when there is a 'perturbation' of our schemata by unexpected experience (shaking the spoon produces no rattle). Such an unexpected experience with the spoon can lead to active adjustments of the learner's schemata through accommodation thereby setting up a new equilibrium. The merits of this as a theory of infant learning will not be questioned here; but its merits as a general theory of acquiring knowledge will be.

Piaget's guiding hypothesis is that our capacity to know *reflects* the autoregulatory mechanisms of the non-cognitive organism. In line with the guiding hypothesis, Piaget extends his theory about the natural processes of accommodation and assimilation of schemata infants use when learning, to all of our cognitive functionings including the acquisition of scientific knowledge and the processes of learning at whatever age. But these natural processes are a far cry from the methods of science that we have self-consciously developed. Piaget's account of the natural processes of infant learning can be viewed as a primitive version of trial and error in which old hypotheses are replaced by new hypotheses when error arises. However even this hardly captures, say, the method of conjectures and refutations, falsification and corroboration, described by Popper, or any other theory of method which tries to capture the idea of 'trial and error'.

Piaget's guiding hypothesis is inadequate in two ways. First, our natural and common-sense modes of reasoning have been shown to be riddled with fallacy by recent research into the psychology of reasoning (e.g., Johnson-Laird & Wason 1977; Kahneman *et al.* 1982; Manktelow & Over 1990; Stich 1994). So the scientific inferences we employ cannot be an extension of, but must correct, the natural processes of reasoning we employ as infants or even adults. This is an important matter that all involved in science education must be wary, particularly constructivists who take all pupil's knowledge to be their own construction. If we are naturally such bad reasoners as research shows, then pupils will need to be corrected time and time again. As the history of science reveals, scientific knowledge has been obtained in the teeth of the errors provided by human 'natural' reasoning about matters. Learners of science are even more prone to repeat such errors.

Second, as the history of philosophy of science shows, there has been a steady development of the methods of science from Aristotle, Bacon, Galileo, Descartes and Newton up to not only Popper but also approaches which rival those of Popper, e.g., probabilists and subjective Bayesians. Importantly there are the special techniques of inference elaborated by logicians, mathematicians and statisticians, e.g., the double-blind randomised tests prescribed by Fisher, which are essential to the ongoing nature of scientific experiment and investigation. It is these methods of reasoning which yield tethering reasons which underpin our knowledge in science. Though Piaget is aware of the complexities of theories of method, his guiding hypothesis misleads us in claiming that these methods are in essence akin to the natural methods infants use to obtain 'knowledge'.

Children at school learning science appear to be in an intermediate position between infants and scientists. But in their learning of science and mathematics they must show some mastery of the reasoning employed by scientists that is more sophisticated than Piaget's theory of the natural processes of assimilation and accommodation to schemata. Our biological evolution may have bequeathed to us as infants certain ways of learning how the world is. But that biological bequest is inadequate for acquiring scientific knowledge. The very methods whereby we obtain scientific knowledge has been improved in the course of the growth of science itself.[14]

3. VON GLASERSFELD AND THE EPISTEMOLOGY OF CONSTRUCTIVISM

Ernst von Glasersfeld is one of the leading proponents of a radical version of constructivism both as a theory of scientific knowledge and as a guide for scientific education. Upon being asked about the difference between radical constructivism (RC) and trivial constructivism (TC) he said:

A few years ago when the term *constructivism* became fashionable and was adopted by people who had no intention of changing their epistemological orientation, I introduced the term trivial constructivism. My intent was to distinguish this fashion from the 'radical' movement that broke with the tradition of cognitive representation (Von Glasersfeld 1993, p. 24).

What is the 'epistemological orientation' that is being rejected?

Constructivism is an attempt to cut loose from the philosophical tradition . . . that knowledge has to be a representation of reality, where reality is spelled with a capital 'R' and what is meant by it is a world prior to having been experienced (*ibid.*, p. 25).

Constructivism does not deal with the traditional concept of 'truth' which would require that one knows or at least believes, that an idea, a theory, or any conceptual construct is an accurate representation or duplication of something beyond the experiential field (*ibid.*, p. 27).

The last remark contains an important error. The traditional concept of truth under attack (i.e., some version of the 'correspondence' theory) does not require that what is true also be known or believed; this is to confuse matters to do with the definition of truth with knowledge or belief about what is true. Setting this error aside we can recognise the sources of Von Glasersfeld's views in Vico and Piaget (amongst others). Thus the first of several theses of Radical Constructivism, RC1, can be stated as follows:

RC1: Truth, and knowledge, are not representations of a reality independent of, or beyond, experience.

As noted in section 1, one way of characterising realism is to say that items of a particular class, e.g., external middle-sized objects, or unobservable entities commonly postulated in science, or numbers, etc., exist independently of the mental activities of people such as perceiving, or thinking about or constructing theories. This is the view of the world commonly endorsed by lay persons and scientists. There would still be things 'out there' even if we were not around to perceive them or to theorize about them. Moreover there are truths about matters other than what we can come to know on the basis of experience; and there are truths we have not yet recognised, and may never do so. It is this commonsense realism about objects and truths that is under attack in RC.

Consider the matter of truth raised in RC1. Do realists need to claim that truths 'represent' reality, whatever this may mean? No. In response consider the seminal account of truth first suggested by Tarski (1949). A necessary (but not sufficient) adequacy condition for any definition of truth is that, for all sentences of a language, the following schema (T) holds:

(T) the English sentence 's' is true if and only if s.

As illustrations consider the following examples of the infinite number of examples of the schema given in the following so-called T-sentences with ordinary sentences in place of 's':

(T1) the English sentence 'snow is white' is true if and only if snow is white;

(T2) the English sentence 'electrons are negatively charged' is true if and only if electrons are negatively charged;

and so on, for all statements expressed in some language.

These examples of T-sentences come close to capturing our ordinary intuitions about truth given by the older correspondence theory which would have said: that snow is white is true if and only if it is a *fact* that snow is white. One of the several advantages of the Tarski approach is that it avoids a commitment to facts in the correspondence theory. Some (e.g., Horwich 1990) would argue that satisfying schema (T) is also sufficient for a definition of truth. This is a minimalist view of truth that also arises from the Tarski approach. The only point to be established here is that, minimally, for truth there is an equivalence, expressed by the words 'if and only if', that holds between the left-hand side of, say, (T1), which is about a sentence, and the right-hand side of (T1), which is about snow and its colour. There is no need to have the stronger notion of sentences *copying*, or accurately *representing*, or *duplicating*, or *picturing* bits of reality that earlier strong versions of the correspondence theory of truth required. This older theory built too much into a theory of truth – hence its replacement by the Tarski theory.

Realists can get by with a quite minimal notion of correspondence as weak as the 1–to–1 pairing suggested by the equivalence relation 'if and only if'. So they can side with Von Glasersfeld's RC1 and reject the strong 'representation' theory of truth while keeping schema (T) as a core requirement for any theory of truth. Trivial constructivists can do the same and also adopt schema (T). However it turns out that Von Glasersfeld's objection to realism turns not primarily on an alleged commitment to a 'representation' theory of truth but on a scepticism about the existence of a reality independent of experience.

Von Glasersfeld does not make the ontological claim that there is no reality beyond experience. Rather he adopts the sceptical epistemological thesis that we cannot *know* anything of a reality beyond experience. All we can know is what is delivered by experience. Thus in answer to the self-posed question 'Does reality exist aside from one's construction of reality?' (*ibid.*, p. 28) the answer is not 'No' but the agnostic 'I do not know'. But note that one must navigate carefully between the two different occurrences of the word 'reality' in the self-posed question. About the referent of the second word of the question, viz., 'reality', we can know nothing; in contrast the last word does refer to something we can know about, presumably the 'reality' we construct out of experience.

We can now formulate the second thesis of RC:

RC2: We can have no knowledge of a reality, prior to, or independent of, experience (including ordinary objects as well as unobservable entities postulated in science).

Note one radical consequence of RC2. It rules out any claim to the effect that scientists have made discoveries about unobservable items such as inertia, energy, electrons, genes, tectonic plates and pulsars. At best we are to remain sceptical of the claim that science has made any discoveries beyond what we can experience. What, then, can we know? '... the only world we can know is the world of our experience' (*ibid.*, p. 24). But in addition; 'Constructivism holds that we can only know what our minds construct...' (*loc. cit*). If we drop the misleading word 'only' from both quotations we have a third thesis that is similar to VEP:

RC3: We can only know either our experiences or what we construct on the basis of these experiences.

However what 'experience' means in RC needs some important clarification.

First, it is important to distinguish the very experiences themselves that we can have from *reports* or *beliefs* about experience; the latter are always involved in knowledge claims because knowledge at least entails belief. Second, it is important to draw a distinction Descartes[15] insisted upon between: (a) knowledge of external things by means of sense-organs as when I say 'I see light', a report which purports to be about an item in the external world, viz., light; (b) the report 'it seems to me that I see a light' which is a description of the sort of experience one is having and not about a beam of light in the external world independent of perceivers. If (b) is the understanding of experience we are to adopt then RC is not the new doctrine it appears to be. It is part of a quite traditional philosophy concerning the existence of the external world that has been one of the main concerns of epistemology from Descartes and the eighteenth-century British Empiricists onwards. Given RC2 and RC3, the charge that RC is a merely a dressed up version of phenomenalism or instrumentalism would be entirely warranted (see Suchting 1992, especially Part I, and Matthews 1994, chapter 7).

Perhaps RC could adopt the more common-sense (a) as its understanding of experience. This would then allow that we can have direct experience, by means of our sense organs, of medium-sized goods, such as tables, chairs and lights, and have no reason to doubt, on the whole, that our experience misleads us about their existence. But this possibility is rejected by von Glasersfeld when he appeals to Einstein's advocacy of *operational definitions* of the terms of science by means of sensory experience, saying: 'Science largely consists of relational (operative) concepts that are the results of various abstractions having their origin in sensory-motor experience and our own mental operations' (*ibid.*, p. 26). Talk of the operational definition of the concepts of science betrays one positivistic tendency in RC that has been known to be quite inadequate for a number of years (for a critical account see Hempel 1965). It turns out that Von Glasersfeld is a thoroughgoing phenomenalist who would not accept (a):

> To conclude that, because we have a perceptual experience which we call a 'chair' there must be a chair in the 'real' world is to commit the realist fallacy. We have no way of knowing what is or could be beyond the experiental interface. If we can reliably repeat the chair experience, we can only conclude that, under the circumstance, it is a viable construct' (Von Glasersfeld 1993, p. 26).

And citing Einstein once more he adds '... our object concepts are "free creations of the human (animal) mind"'. In claiming that ordinary objects are merely our viable constructions, von Glasersfeld becomes as radical a philosophical idealist as Bishop Berkeley.

In RC is experience something of which we are passive recipients? Or are we active as knowers even when we have experience? If strong emphasis is placed on the active role of construction not only of ordinary objects but also of experience itself then it is possible to attribute to RC the thesis, advocated by Hanson (1965), Kuhn (1970) and many others, that all observation is theory-laden: any experience a person has is laden with the conceptual structures each has constructed.

The conceptual structures embodied in our scientific theories and hypotheses carry with them a commitment to certain kinds of entities. However in the light of theses RC1 to RC3 we can never *know* whether reality actually contains the entities postulated in our theories. It is *as if* we live in a world populated by what our conceptual structures posit. Our conceptual structures not only populate the theoretical reaches of science with entities, but they go right down the line to shape our very experiences and reports of experience that have been commonly assumed to be free of any taint by theory. What, then, constrains our adoption of the conceptual structures that penetrate experience? Independent reality can not play this role since we can know nothing of it; and nor can experience since it too is riddled with theory. We thus end up in a radical relativism about both what exists and what experience there is.[16] This radical relativism is not something to which empiricists such as Berkeley, Hume or the positivists are committed; they alleged that experience is the solid rock foundation upon which all knowledge is to be built. Thus a further thesis which takes RC beyond empiricism to relativism is:

> RC4: What exists and what 'truths' we may say there are, are relative to those conceptual structures that each of us constructs.

The relativism of RC4 comes close to a solipsism of the individual constructor in his or her own world of personal constructs. Von Glasersfeld envisages that someone who grasps the solipsism implicit in his position might say: if each of us is locked into our own experiential world, how can we agree on anything or communicate with one another? His response is to say that there is agreement and communication. But we do not explain the possibility of agreement and communication in terms of either the existence of shared experiences, or common objects of which we have experience, or a shared reality or shared meanings. We explain it as follows:

> If two people or even a whole society of people look through distorting lenses and agree on what they see, this does not make what they see any more *real* – it merely means that on the basis of such agreements they can build up a consensus in certain areas of their subjective worlds. Such areas of relative agreement are called 'consensual domains'... (Von Glasersfeld 1991, pp. xv-xvi).

It is important to appreciate the radical character of RC's account of science and knowledge embodied in the above remark. Realists, and perhaps trivial but not radical constructivists, claim that scientific knowledge is largely constrained by intersubjective observations by scientists of an independent reality which, as we discussed in section 2.3, can say either 'Yes' or 'No' to the theories we propose. But for Von Glasersfeld each of us only has access to our own experiences; we can have no access to, and must remain sceptical about, any intersubjective observations or any independent reality that our theories may allegedly be about. Neither observation nor reality play any role in RCs' account of science. All that constrains us in our scientific 'knowledge' is our social interaction with others and the adaptation that we make with one another. Summing this up as a further thesis we have:

RC5: Testing of hypotheses in science is a matter of consensus or dissensus within the community of scientists and not a matter of the 'Yes' or 'No' of nature.

This position results from the abandonment of any role for truth and an agnosticism about an independent world to which our beliefs might respond, leading to a scepticism about the existence of anything beyond one's immediate personal experience. All that one might do in forming beliefs is reach a consensus with other believers. But given such a radical scepticism how can we even be sure that there are other believers, let alone form any kind of consensus with them?

If we simply assume realism with respect to other people (i.e., we do not construct the other out of our own experience), then little is said about how or why we may be so lucky as to even agree amongst one another, or even how we may set up intercommunal signs for agreement and disagreement in the first place. But we are told:

> The consensual domain of the scientific subculture, because of its many specific constraints, is particularly homogeneous, and new hypotheses are 'tested' against that (relatively) homogeneous background (Von Glasersfeld 1993, p. 28).

That the word 'tested' appears in inverted commas is no accident because it can no longer mean what it normally means. Our scientific beliefs are no longer tested against the deliverances of intersubjective observations, or by the 'Yes' and 'No' of independent reality. In fact they are not tested at all. We accept what science says because accepting it makes the least perturbations with what others accept and, in the long run, with what the culture of scientists accepts. The scientific 'knowledge' we have results only from negotiated consensus. Von Glasersfeld has thus fallen in with constructivists in the sociology of scientific knowledge.

The radical position just outlined appears to be in tension with a more conservative, but still implausible, position when Von Glasersfeld says of RC: 'its constructing is not free: ... some of our conceptual constructs work and others do not' (*ibid.*, pp. 24–5). And as already cited:

> As for Vico, knowledge for Piaget is never (and can never be) a 'representation' of the real world. Instead it is the collection of conceptual structures that turn out to be adapted or, as I would say, *viable* within the knowing subject's range of experience (Von Glasersfeld 1989, p. 125).

It appears after all that our conceptual structures are to be judged as viable (adapted) or not viable (ill-adapted) with respect to experience. But if experience is to come into conflict with a conceptual structure then it can not be laden with *that very* conceptual structure, i.e., the doctrine of the theory-ladenness of experience must either be understood in a quite conservative way or abandoned.[17] The more conservative position can be expressed:

> RC6: Conceptual structures are either viable or not viable with respect to experience; they are neither true nor false.

Viability, as a way of evaluating conceptual structures, is said to be different from truth. But about prediction we are told: 'If a prediction [in science or in life] turns out to be right, a constructivist can only say that the knowledge from which the prediction was derived proved viable under the particular circumstances of the case' (Von Glasersfeld 1993, p. 26). Despite disclaimers, RC cannot escape truth. Even if RC rules out talk of truth concerning the non-experiential reaches of our conceptual structures, what does it say about predictions concerning what we might experience? Note that in the last quotation Von Glasersfeld envisages what happens if 'a prediction turns out to be *right*'. But talk of 'right' is simply to ask whether an observational consequence of a theory is true (or false) when compared with what we observe on the basis of experience. Thus talk of the viability or unviability of conceptual structures is merely a mask for talk of truth or falsity of predictions when matched against observation.

Von Glasersfeld says nothing about what happens when a prediction goes wrong (or a sufficient number go wrong). Presumably he would say that the conceptual structure from which the prediction was derived is not viable in those circumstances. But once more this masks underlying talk of falsification. Clearly some elementary methodology of confirmation and falsification must be provided if we are to move beyond the rather vague claim that conceptual structures are either viable or unviable.

We should bear in mind that for Von Glasersfeld experience, and the conceptual structures adapted to experience, are those of the individual person. They are not the intersubjective observations and the theories held in common with which realists assume the community of scientists deal. For von Glasersfeld's scientists, any communality of theory results from consensus about what each finds viable within each person's own world of experience. The common reality about which some scientists think they are making discoveries is an illusion, given the radical character of RC. But matters are even more radical than this since not only are there no shared experiences. There are no shared meanings:

> RC7: Just as there is no common experience for each scientist, so there are no shared meanings amongst the community of scientists concerning the theories they construct.

Absence of shared meanings is not merely a matter of failing to grasp what another means, something that a prolonged discussion could overcome. Rather, even with the best will in the world one person cannot mean what another means: 'Your meaning and another's are at best compatible; in a given situation neither reacts in a way that the other could not expect' (*ibid.*, p. 32; also p. 35). There is no shared meaning, or sameness of meaning, but only consensus about meaning, a sign of which is compatibility. But 'compatibility' remains unexplained. Incompatibility requires some sameness of meaning as when one asserts 'the litmus is red' while another asserts 'the litmus is not red'. Here words mean the same, but one denies what the other asserts. Disagreement only takes place against a background of shared meaning. If there is no shared meaning then trivial compatibility arises; we discourse on different topics and mean quite different things thereby constantly talking past one another. If we do not share meanings then all our beliefs are liable to be about quite different matters, a consequence which undermines the possibility of common scientific knowledge of how the world is. There is an even more radical incommensurability of meaning than that advocated in Kuhn (1970). At least for Kuhn a community of scientists can share a paradigm (i.e., disciplinary matrix) incommensurable with the paradigm of another community; for Von Glasersfeld each scientist lives within their own edifice of incommensurable self-constructed meanings.

What of knowledge? We may say crisply:

> RC8: All knowledge is construction.

But such alleged 'knowledge' lacks Plato's tether of reasons as when we are told: 'Knowledge is often built up by combining and recombining available concepts or by trying out new conceptual relations' (*ibid.*, p. 34). Apparently no reasons are required for accepting one combination of concepts rather than some other as 'knowledge'. Nor will it do to tell us that 'the test for knowledge is not whether or not it accurately matches the world ... but whether or not it *fits* the pursuit of our goals ... within the confines of our own experiential world' (Von Glasersfeld 1990, p. 31). However to say that a test for knowledge of, say, a hypothesis is that it fits experience is to commit the fallacy of affirming the consequent. For this reason the 'viability' of our conceptual structures with experience must be just that; nothing is to be inferred about the truth, falsity, probability or otherwise of our hypotheses, something that RC2 and RC3 underline.

Elsewhere false oppositions are set up about knowledge between constructivists and non-constructivists as when we are told: 'Knowledge is always the result of a constructive activity and, therefore, it cannot be transferred to a passive receiver. It has to be actively built up by each individual knower' (Von Glasersfeld 1993 p. 26). As we have seen the Socratic teacher requires the active construction of reasons for beliefs by the individual knower (section 2.1); and the scientist is active in conjecturing theories and looking for test evidence (section 2.3). Opponents of RC can admit that there is an active component to knowing while denying that knowledge itself is merely construction.

In sum, RC is an unstable mixture of anti-realist doctrines from empiricsm, scepticism, solipsism, relativism and the sociology of scientific knowledge mixed with a dollop of anti-rationality.

4. A SUMMARY OF OBJECTIONS TO CONSTRUCTIVISM IN SCIENCE AND SCIENCE EDUCATION

The previous section is largely a critique of the conception of science and of knowledge found in RC. Little is said about science education itself. What then can be said of a science education which takes its cues from RC's conception of knowledge? Opponents of RC could happily admit that the metaphor of construction can be extended to methods of teaching and learning, as the episode with the Slave-Boy of section 2.1 shows. And they can readily accept new proposals for the empirical study of methods of efficacious teaching proposed by constructivists. Thus Von Glasersfeld goes beyond the activities of a Socratic teacher when he suggests: 'Teachers have a better chance to modify a student's conceptual structures and understanding if their interventions are informed by a hypothetical model of the student's present ideas' (Von Glasersfeld 1989, p. xix). Again opponents of RC could happily extend the metaphor of construction to the activities of scientists when they invent hypotheses and look for methods of testing them, as discussed in section 2.3. However none of these extensions of the metaphor of construction to student teaching and learning or to scientists' invention bear on the question 'What is scientific knowledge?'

In the light of what has been said so far at least eight objections can be raised to constructivism as an account of knowledge in science education.

Objection 1. Radical constructivism in science education is partisan and not neutral about the nature of scientific theories and knowledge and science education.

The dispute between the objectivist and the non-objectivist interpretation of scientific theories is philosophical in character. Theories within science education are powerless to resolve the dispute. By taking one side rather than another a constructivist science education adopts a partisan and non-neutral stand on the issue. This leads to my second objection.

Objection 2. Radical constructivism in science education presents a distorted picture of the history and successes of science.

Throughout the history of science scientists have proposed theories which they understood realistically or non-realistically. If the matter of an objectivist versus non-objectivist understanding of science were to arise in the course of teaching science then a science education conducted from a purely constructivist point of view would not do justice to the history of science. In fact since most scientists understand their science in an objectivist way, constructivism would provide a distorted picture not only of the history of science but also contemporary science. This leads to the next objection.

Objection 3. Radical constructivism is impotent to convey to anyone the body of well established theoretical truth that exists in science.

Since radical constructivists deny that there is any theoretical truth then, unless they wish to seriously mislead students, there is much established science that they would have to deny, in virtue of their account of 'scientific knowledge'. As we have seen, and despite some equivocation on the matter, they would have to deny not only knowledge of, but the very existence of, such things as gravitational attraction, genes, viruses, and so on. They would have to represent all of, say, our theories of diseases caused by viruses as mere constructs which are viable with our experience of being ill or healthy.

Objection 4. Radical constructivism depends on a discredited philosophy of science.

RC1 presents a misleading account of truth. If philosophy of the past 40 years has shown anything it is that the phenomenalistic idealism underlying RC2 and RC3 is wrong. Though the message has taken a long time to filter down, doctrines such as RC4 about the relativity of truth and existence have been in retreat since Plato's criticism of Protagoras in his *Theaetetus*. And RC6 turns on a lamentably weak account of scientific method.[18] Thus a science education modelled on RC would carry the burden of discredited philosophical doctrines.

The fifth objection can be developed as follows. We need to keep separate constructivism as an interpretation of scientific theories and constructivism as a theory about the way in which students come to learn about science. These are two quite distinct matters that von Glasersfeld runs together when he talks of 'the conceptual structures that epistemic agents, given their range of experience... consider viable' (Von Glasersfeld 1989, p. 124). But which epistemic agents? *Scientists*, who invent theories? Or *pupils* who are to be taught, or learn, about the scientific theories that scientists have invented? If constructivism is 'viable' for the learning of scientific theories by students, it does not follow that the theory learned has to be understood constructivistically; or conversely. The scientists' interpretation of his theory, whether constructivist or objectivist, is quite different from the student's learning, whether constructivist or not, of the scientists' theory. The objection may be summarised thus:

Objection 5. The radical constructivist interpretation of scientific theories does not ground, and is quite independent of, the constructive activities of students who learn scientific theories; i.e., the constructivist account of the learning or teaching of science has nothing to do with the constructivist interpretation of the science learned.

Objection 6. Radical constructivists with their talk of 'viability' and their relativist inclinations have lost the idea of a right and a wrong answer in science and science education.

This is certainly the case if each of us is a solipsistic constructor of our own conceptual structures out of our own experiences of the sort depicted by RC5 and RC7. In response to the question 'What harm is there in remaining a realist?' Von Glasersfeld replies: 'No harm at all, as long as you don't tell others that the reality you have constructed is the one they ought to, or worse, must believe in' (Von Glasersfeld 1993, p. 28). In one sense, one could agree that merely telling someone that such-and-such a theory is the correct one to believe is not a good way to teach the theory. However constructivists ignore the fact that there are correct theories and *that* a theory is correct can be a spur to someone to start learning the theory. However behind this remark is the false belief that objectivists claim that they know *with absolutely certainty* that their theories are true. No objectivist this century has held this view. As Popper, a leading objectivist, has pointed out time and time again, all our theories are fallible; the theories we accept may be (partially) true; but we cannot show this with the highest degree of certainty.

Objection 7. Objectivism adopts an authoritarian stance in science education and encourages passive learning while constructivism does the opposite.

This charge often follows from the false claim that objectivists know the truth of scientific theories with absolute certainty; this is turned into a reason for rejecting objectivism because it is authoritarian. But worse, it is alleged that objectivist views about knowledge always lead to didactic teaching:

> Objectivism [is] . . . the foundation of the traditional model of education. In this model it is assumed that an already developed body of knowledge, developed, proven, and accepted by society, can easily be transmitted to students through generally passive instructional means. . . . No wonder the process of teaching is defined to be telling and the process of learning is often considered to be memorization and recall (Dana & Davis 1993, p. 326).

It can be agreed that there are those who take an objectivist view of scientific knowledge and who are didactic teachers. But must an objectivist view of scientific knowledge always lead inevitably to didactic teaching? Clearly the above passage assumes this; so if one is to get away from didacticism one must abandon the objectivism that leads to it. Once freed from objectivism it is alleged that other modes of teaching become available, or as Dana and Davis go on to say: 'if another epistemological perspective becomes the referent, these all-too-familiar conceptions of knowledge, knowing, curriculum, and instruction all come into question'. However the whole point of section 2.1 was to show, through the example of Socrates, that one can be an objectivist about knowledge without adopting a didactic view of teaching. In throwing out the dirty bath water of didacticism constructivists also throw out the baby of objective knowledge.

The final objection is:

Objection 8. Constructivists have a distorted conception of knowledge which infects their view of learning.

That this is so is the main point of this paper.

Not all constructivists may wish to be saddled with the extreme position of RC. They may wish to salvage some of its less radical aspects. Such appears to be the case in the 'position paper' of Driver *et al.* (1994). They begin promisingly enough with the words: 'The core commitment of a constructivist position, that knowledge is not transmitted directly from one knower to another but is actively built up by the learner, is shared by a wide range of different research traditions . . . .' (Driver *et al.* 1994, p. 5). A Socrates would agree, and might include himself amongst those traditions – until he reads further down the page and finds that 'scientific knowledge is both symbolic and socially negotiated'. Talk of 'social negotiation' is part of the vocabulary of constructivist theories of the sociology of scientific knowledge in which the role of reasons and reasoning in knowledge is thoroughly downplayed in favour of negotiations over hypotheses. So the account given by Driver *et al.* of scientific knowledge cannot be the account of knowledge of Socrates outlined in section 2.1.

After the last remark the authors go on to say: 'The objects of science are not the phenomena of nature but constructs that are advanced by the scientific community to interpret nature?' In regarding objects as constructs, Driver *et al.* make a point about the nature of scientific theory and not about education. Are we to understand their remark in von Glasersfeld fashion, viz., as a denial that scientific theories are ever about an independent reality? This would be a misreading because, despite their use of non-realist vocabulary, the authors endorse Harré's (1986) realism saying: 'In proposing a realist ontology, Harré (1986) suggests that scientific knowledge is constrained by how the world is and that scientific progress has an empirical basis, even though it is socially constructed and validated' (*ibid.*, p. 6).

Various elements sit uneasily here. On the one hand science is constrained by how the world is, but on the other hand objects are constructs. Clearly we are owed an account of when the objects postulated in science are found to be real but unobservable objects in the world. This is important if science is to be taught as something which has made important discoveries about the world and how it works. On this aspect of teaching constructivism must remain silent. Again we are told that science has an empirical basis, but that in the long run the sciences we adopt are 'socially negotiated'. Nothing is said about how the 'empirical basis' counts as evidence for a theory; it would appear it counts not at all since negotiation plays the role of adjudicator of theories. Once more the role of reasons in knowledge and our sciences is quite enfeebled.

Driver *et al.* are mentioned here only as an example of those who may wish to distance themselves from some of the more radical constructivist doctrines of von Glasersfeld. But they have still not freed themselves from its most central doctrine, viz., the constructivist account of knowledge. What is needed in science education is renewal of epistemology, both in respect of its account of science itself and in its account of learning and teaching. It seems with respect to epistemology that many of the recent moves in constructivist science education have been steps backwards, not forwards, from Socrates and Plato.

NOTES

1. K. Tobin (ed.), 1993, 'Preface', p. ix.
2. Many educationalists are unmindful of the fact that Thomas Kuhn, who introduced the term 'paradigm' in the 1962 edition of his *The Structure of Scientific Revolutions* effectively abandoned it in the 'Postscript' to the 1970 second edition. He recommends that the term 'paradigm' be replaced by 'disciplinary matrix' (p. 182) and 'exemplar' (p. 187). In his post-*Structure* historical work Kuhn has avoided the use of the term 'paradigm' and in a recent interview declares that 'I'm not sure I would use the term paradigm in such a wide sense any more' (Borradori 1994, p. 166).
3. Tobin (ed.), *ibid.*, p. ix.
4. On the difference between metaphysical realism and a realism defined in terms of truth see Devitt (1991, Part II).
5. On idealisation see Cartwright (1983); on approximation to the truth or verisimilitude see Popper (1963, chapter 10 and Addenda 6 and 7).
6. See Van Fraassen (1980, chapter 2, section 1). For an evaluation of Van Fraassen's position see the papers in Churchland & Hooker (eds.) (1985). For a more general criticism of constructivism see Boyd (1992) and Rosen (1994).
7. For more on this theme see Siegel (1993).
8. Note that the theory of knowledge about the Forms made famous in Plato's *Republic* is quite distinct from, and irrelevant to, the theory developed in the *Meno* and the *Theaetetus*.

9. This is especially the case since the objections raised by Edmund Gettier to the definition given. For a review of the subsequent literature see Shope (1983) and Plantinga (1993).
10. 'Vico's Concept of Knowledge' in Berlin (1981).
11. Popper (1959, sections 1–3) and Reichenbach (1938, section 1).
12. Popper (1972, p. 24; see also p. 68 and p. 328). See also Popper (1963, p. 117).
13. On the question 'why aim for knowledge rather than true belief?' see Dretske (1989).
14. Other important philosophical influences upon constructivism cannot be mentioned here; further influences are mentioned in Hawkins (1994).
15. Descartes' 'Second Meditation' of his *Meditations*.
16. For two critical responses to these aspects of the work of Kuhn, Hanson, Feyerabend and to constructivism more generally see Scheffler (1967) and Boyd (1992).
17. See Scheffler (1967, especially chapter 2), for a critical account of the way in which radical theory-ladenness undercuts objectivity.
18. Criticisms of the above theses are common throughout present-day philosophy and only some can be cited here, e.g., Boyd (1992), Devitt (1991), Nola (1994), Scheffler (1967), and Stove (1991, chapters 5 and 6). The constructivist views of Van Fraassen (1980), are quite different from those of Von Glasersfeld and are not to be countered in the same way; for a critical account see Churchland & Hooker (1950), Devitt (1991, chapter 8) and Rosen (1994).

REFERENCES

Berlin, I.: 1981, *Against the Current: Essays in the History of Ideas*, Oxford University Press, Oxford.
Borradori, G.: 1994, *The American Philosopher: Conversations with Quine, Davidson, Putnam, Nozick, Danto, Rorty, Cavell, MacIntyre and Kuhn*, The University of Chicago Press, Chicago.
Boyd, R.: 1992, 'Constructivism, Realism and Philosophical Method', in J. Earman (ed.), *Inference, Explanation and other Frustrations*, University of California Press, Berkeley, pp. 131–198.
Cartwright, N.: 1983, *How the Laws of Physics Lie*, Clarendon Press, Oxford.
Churchland, P. M. & Hooker, C. A. (eds.): (1985), *Images of Science*, University of Chicago Press, Chicago.
Dana T. M. & Davis, N. T.: 1993, 'On Considering Constructivism for Improving Mathematics and Science Teaching and Learning', in K. Tobin (ed.), *The Practice of Constructivism in Science Education*, AAA Press, Washington, D.C., pp. 325–333.
Devitt, M.: 1991, *Realism and Truth*, second edition, Blackwell, Oxford.
Dretske, F.: 1989, 'The Need to Know', in M. Clay & K. Lehrer (eds.), *Knowledge and Skepticism*, Westview Press, Boulder, pp. 89–100.
Driver, R., Asoko, H., Leach, S., Mortimer, E. & Scott, P.: 1994, 'Constructing Scientific Knowledge in the Classroom', *Educational Researcher* 23, 5–12.
Hanson, N. R.: 1965, *Patterns of Discovery*, Cambridge University Press, Cambridge.
Harré, R.: 1986, *Varieties of Realism*, Blackwell, Oxford.
Hawkins, D.: 1994, 'Constructivism: Some History', in P. J. Fensham, R. Gunstone & R. White (eds.), *The Content of Science: A Constructivist Approach to Teaching and Learning*, The Falmer Press, London, pp. 9–13.
Hempel, C. G.: 1965, 'A Logical Appraisal of Operationism', *Aspects of Scientific Explanation*, Free Press, New York, pp. 123–133.
Horwich, P.: 1990, *Truth*, Blackwell, Oxford.
Johnson-Laird, P. N. & Wason, P. C. (eds.): 1977, *Thinking: Readings in Cognitive Science*, Cambridge University Press, Cambridge.
Kahneman, D., Slovic P. & Tversky A. (eds.): 1982, *Judgment Under Uncertainty: Heuristics and Biases*, Cambridge University Press, Cambridge.

Kant, I.: 1934, *Critique of Pure Reason*, J. M. D. Meiklejohn (trans.), Dent, London.
Kuhn, T. S.: 1970, *The Structure of Scientific Revolutions*, second edition, The University of Chicago Press, Chicago.
Manktelow, K. I. & Over, D. E.: 1990 *Inference and Understanding: A Philosophical and Psychological Perspective*, Routledge, London.
Matthews, M. R.: 1994, *Science Teaching: The Role of History and Philosophy of Science*, Routledge, New York.
Nola, R.: 1994, 'There are More Things in Heaven and Earth, Horatio, Than are Dreamt of in Your Philosophy: A Dialogue on Realism and Constructivism', *Studies in History and Philosophy of Science* 25(5), 689–727.
Piaget, J.: 1971, *Biology and Knowledge*, Edinburgh University Press, Edinburgh.
Plantinga, A.: 1993, *Warrant: The Current Debate*, Oxford University Press, Oxford.
Plato: 1956, *Meno*, in *Protagoras and Meno*, W. K. C. Guthrie (trans.), Penguin, Harmondsworth.
Popper K. R.: 1959, *The Logic of Scientific Discovery*, Hutchinson, London.
Popper, K. R.: 1963, *Conjectures and Refutations*, Routledge and Kegan Paul, London.
Popper, K.. R.: 1972, *Objective Knowledge*, Clarendon, Oxford.
Reichenbach, H.: 1938, *Experience and Prediction*, The University of Chicago Press, Chicago.
Rosen, G.: 1994, 'What is Constructive Empiricism?', *Philosophical Studies* 74, 143–178.
Scheffler, I.: 1967, *Science and Subjectivity*, Bobbs-Merrill, Indianapolis.
Siegel, H.: 1993, 'Naturalized Philosophy of Science and Natural Science Education', *Science & Education* 2(1), 57–68.
Shope, R. K.: 1983, *The Analysis of Knowing*, Princeton University Press, Princeton.
Stich, S. P.: 1994, 'Could Man be an Irrational Animal? Some Notes on the Epistemology of Rationality', in H. Kornblith (ed.), *Naturalizing Epistemology*, second edition, MIT Press, Cambridge, MA, pp. 337–357.
Stove, D.: 1991, *The Plato Cult and other Philosophical Follies*, Blackwell, Oxford.
Suchting, W. A.: 1992, 'Constructivism Deconstructed', *Science and Education* 1, 223–254.
Tarski, A.: 1949, 'The Semantic Concept of Truth', reprinted in Feigl H. & Sellars W. (eds.), *Readings in Philosophical Analysis*, Appleton-Century-Crofts, New York, pp. 52–84.
Tobin, K. (ed.): 1993, *The Practice of Constructivism in Science Education*, AAA Press, Washington, D.C.
Van Fraassen, B. C.: 1980, *The Scientific Image*, Clarendon, Oxford.
Von Glasersfeld, E.: 1989, 'Cognition, Construction of Knowledge and Teaching', *Synthese* 80, 121–140.
Von Glasersfeld, E. (ed.): 1990, 'Environment and Communication', in L. P. Steffe & T. Wood (eds.), *Transforming Children's Mathematical Education*, Lawrence Erlbaum Associates, Hillsdale, NJ, pp. 30–38.
Von Glasersfeld, E. (ed.): 1991, 'Introduction', *Radical Constructivism in Mathematics Education*, Kluwer Academic Publishers, Dordrecht, pp. xiii–xx.
Von Glasersfeld, E. (ed.): 1993, 'Questions and Answers about Radical Constructivism' in K. Tobin (ed.), *The Practice of Constructivism in Science Education*, AAA Press, Washington, D.C., pp. 23–38.

Constructivism Deconstructed

W. A. SUCHTING

231 Bulwara Rd, Ultimo, NSW 2007, Australia

> ... *ridentem dicere verum*
> *Quid vetat?*
> Horace, *Sat.*, I.i.24

ABSTRACT: This paper examines the doctrine of 'constructivism' as presented by Ernst von Glasersfeld (1989). Part I attempts to elicit a clearer statement of the concepts, positions and arguments for the latter than is immediately available in the paper. Part II discusses the problem of intersubjectivity in constructivism. The general conclusions drawn from these sections is that the basic concepts and theses of constructivism are, mostly, at best very obscure, that there is very little argument involved, and that where there is it is quite unsatisfactory. Part III ventures an explanation of at least some of the weaknesses in the doctrine, this involving a brief independent treatment of some relevant epistemological questions.

1. This paper is concerned with a doctrine which has for some time been very influential in thinking about education, namely 'constructivism', associated especially with the name of its originator and principal exponent, Ernst von Glasersfeld.[1] More specifically, it will examine the doctrine as presented in that author's 'Cognition, Construction of Knowledge, and Teaching' (1989). This has been chosen as a textual basis for the discussion not only because it is by von Glasersfeld himself, but also because it is sufficiently recent for the reasonable assumption to be made that it contains his latest public thoughts on the theory; furthermore, it is brief enough to be considered comprehensively but compendiously. All otherwise unattributed quotations and page references in the main body of this paper will be from and to this publication.

I

> And all were amazed and perplexed,
> saying to one another, 'What does this mean?'
> *Acts* 2:12

2. The first task is to identify and to state as clearly as possible for present purposes the principal concepts, positions, and supporting arguments distinctive of constructivism. This doctrine itself holds that 'language users must individually *construct* the meaning of words, phrases, sentences and

texts' (132), so we must expect, if this is so, to have to do the same for that theory itself.

If we focus first on basic positions, which of course presuppose basic concepts, then we may remind ourselves that there are in general two ways of trying to specify the fundamental theses of a doctrine. One is direct, an attempt to state them just as such. The other is indirect, following the lead of Spinoza's dictum *determinatio negatio est* (Letter 50), to see what is being affirmed by trying to see what is being denied, excluded; this is often illuminating, especially when the theses offer some obstacles to being identified. This approach, to the extent that it is successful, will yield at least a delimited range of possibilities as to what is actually being asserted. Adapting a theological distinction and terminology, the first method may be called the *via affirmativa* and the second the *via negativa*.[2] Our beginning here with the latter is suggested by the fact that the paper being discussed commences with a reference to what constructivism opposes, the 'theory of knowing' to which it is an 'alternative' (121). So this is where, in fact, a start will be made.

3. The paper opens by saying that what constructivism denies is 'the existence of objective knowledge and the possibility of communicating it by means of language' (121).

3.1. Firstly, what is the point of the last four words? Taken by themselves they would naturally be understood to state or imply a qualification of some sort. But there is no hint as what this might be intended to be, and it seems impossible to see how 'objective knowledge', were it to exist, could be communicated *except* by means of a language of some sort. So there is nothing for it but to consider them otiose and in effect to elide them.

Secondly, consider the five words immediately preceding those just commented upon. They are also very puzzling in the context of the cited passage in which they appear. For if 'it', that is, 'objective knowledge', is asserted not to exist, and does not exist, there can be no question of the possibility of communicating 'it' (linguistically or in any other way), as distinct from the possibility of merely *seeming* to communicate 'it'. (It is rather like saying that God does not exist and we cannot communicate with Him either.) Could it be that the non-existence of 'objective knowledge' is asserted only as a contingent matter, and that what is meant is that if such knowledge were to exist, contingently, then it could not be communicated (contingently or necessarily, as the case may be)? But since it is said elsewhere (e.g. 135) that 'objective knowledge' is *impossible*, there can be no question about what might be the case were it to exist. So it would seem as though this part of the text can be excised, too.

3.2. What is left is the denial of 'the existence of objective knowledge'. Now, since constructivism certainly wants to affirm the existence of knowledge in *some* sense the key word here must be 'objective'. What is meant by this in the present context? The paper says at one point, in passing (138, note 1), that the word is commonly used in two ways: '(a) referring to knowledge that purports to describe the world as it is; and (b) knowledge that purports to be intersubjective.' Now this cannot be quite right, for what is meant by qualifying knowledge as 'objective' must surely be not that it merely 'purports' to describe the world as it is or to be intersubjective, but that it *does* describe the world as it is, or *is* intersubjective, though of course that claim may be incorrect, in which case the knowledge would indeed only 'purport' to be thus descriptive or intersubjective. So what must be meant is that 'objective' is commonly used to claim (a) their knowledge describes the world as it is and/or (b) their knowledge is intersubjective. Now, since constructivism itself wants to claim (with what right there will be later occasion to inquire) that knowledge is intersubjective, sense (b) of 'objective' can be ignored here, and attention can be directed exclusively to (a). So, by simply replacing *definiendum* by *definiens* in the expression cited at the beginning of this paragraph, what is denied by constructivism becomes: 'the existence of knowledge that describes the world as it is'.

3.3. There are at least two problematic items here, namely, 'describes' and 'as it is'.

3.31. What is the import here of 'describes'? From the evidence of various passages (e.g. 122, 135) it would seem to be used more or less synonymously with 'represents', for 'representation' is the word used to refer to what is being rejected at those places. (The reprobation of 'representation' is sometimes marked by enclosing the word in scare quotes.) But what is it about what is signified by these terms that is being rejected? This is never stated. But the chief textual clue to 'constructing' what it is may well be the close association of both with 'corresponds'. This is presumably meant in the manner in which the cognate occurs in the 'correspondence theory of truth'. This may be said to affirm that the truth of propositions is determined by some one–one correspondence between the terms of the proposition and the elements of some fact, where 'correspondence' means (briefly, and therefore roughly) that there is some common relation between each member of the one class and some member of the other, the relation being customarily left fairly vague though something like a 'picturing' relation (cf. Wittgenstein's *Tractatus*) seems most often to be meant. So the formulation at the end of 3.2 may now be restated as: 'the existence of knowledge that corresponds to the world as it is', understanding 'corresponds' in the way just explained.

3.32. What, now, is the import of '[the world] as it is'? At another place this seems to be used synonymously with 'what exists' (135). The problem with expressions such as these in the context of what is being denied is that, taken by themselves, they are vacuous. For anything at all that purports to be knowledge must, in some sense or other, be about the world 'as it is', or 'what exists'. The point here is: what, more exactly, *is* it that 'is' or 'exists'? (The case is similar with a claim like: 'The future will be what it will be', which has sometimes been thought to express some particular docrine, usually fatalism, when it is really just a dry pleonasm from which no philosophical milk can be squeezed.) The connection of these locutions with 'real' (125) is by itself not of much help. (The cognate 'reality' is sometimes scare-quoted – for example, 121 – sometimes not.) The scent gets stronger with phrases like 'ontological reality', 'absolute reality' (129,135). In particular, since 'absolute' means, among other things, 'free of all dependence', 'totally independent', we may take it that what all these locutions point to is what is termed at once place (122) 'an observer-independent world-in-itself'. So, putting together all the various parts of the puzzle such as have been assembled so far, we have that what constructivism denies is 'the existence of knowledge that corresponds to an observer-independent world-in-itself'.

3.4. Before beginning the next stage of the exegetical journey, this time on the highway of the *via affirmativa*, it may be useful to issue a caution about a false trail. Here and there along the *via negativa* there occurs what may be taken to be a signpost: a rejection of 'the quest for immutable objective truth' (122). The association here of 'immutable' with 'objective' as qualifying 'truth' might well suggest to the hasty and incautious that the latter is a sufficient condition for the former, or even a necessary condition. This may be reinforced by talk of denial of 'certain knowledge about the world – objective knowledge' (135). But in fact 'immutability' or 'certainty' has nothing essential to do with the question of 'objectivity', in the sense in question, when all are used in the context of questions about knowledge. For example, the so-called 'Galilean' transformation equations of classical kinematics proved not to be 'immutable', insofar as they are replaced in special relativity by different, more general equations. But the former were not simply rejected, but shown to be only conditionally applicable, the limits of applicability being exactly specifiable. So the approximations the Galilean equations permit the physicist to calculate are not less 'objective' than the previous, putatively non-approximative ones. As regards 'certainty' it is perfectly proper to say (within this whole epistemological framework) that the statement: 'Isaac Newton, the famous physicist, was born on 4 January 1643' is true and an instance of 'objective' knowledge, and *also* that it is not an instance of certain, that is, incorrigible knowledge (if such there be). This is so if only because statements of that *sort* are sometimes false (even if believed to be true), and, even if this were not the case, such statements are such that it is easy to see how the evidence for their truth might be less than absolutely probative. So let this linking of 'immutability' and 'certainty' with the idea of 'objectivity' regarding knowledge be set aside once and for all.

4. To approach the problem of presentation of constructivism more positively now, it may be said to begin with that if it denies the thesis formulated at the end of 3.32 above, then, assuming that it affirms the existence of knowledge of an alternative sort, the latter must be knowledge of a world that is *not* observer-independent, that is, it must be knowledge of an observer-dependent world. This position, so far inferred from what is said to be denied, is in fact also to be gathered from various things that

are said on the second page (122) of the paper under examination. Thus it is said here that 'the conceptual constructs we call knowledge' have their place in 'the experiential world of the knowing subject', in 'subjective realities', 'biological organisms' self-generated environments', the individual's subjective reality', which is a '*construction*'. A number of different, even if related things are said here and it will be well to try to get them into focus. The following is offered as at least a first list of main theses.
1. Knowledge consists of 'conceptual constructs'.
2. Knowledge relates to (subjective) experience.
3. This experience is that of *individual* subjects.
4. This individual experience is 'reality' for the subject of the experience.
5. This reality is a 'construction', 'self-generated'.
Now, with regard to (1), whilst being a 'conceptual construct' may well be a necessary condition for some item's being an instance of knowledge, it cannot also be a sufficient condition, for some 'conceptual constructs' are not instances of knowledge. In fact, the paper tells us that for such to be instances of knowledge they must be 'viable' in or for experience (122, 124,125, etc.), and later on (135) that knowledge has to be what is 'feasible' in experience, 'viable' and 'feasible' being thus apparently synonymous. Furthermore, it is said (135) that knowledge *qua* 'feasible' is a 'mapping'. So 'viable'/'feasible' would seem to be what replaces, for constructivism, objectivism's 'paradoxical conception of truth that requires a forever unattainable ontological test' (129), and 'mapping' is what replaces 'description'/'representation'/'correspondence'. So (1) above should be reformulated, and this may be done as follows.

(1'a) Knowledge
    (1'a) consists of 'conceptual constructs', that
    (1'b) are 'viable'/'feasible' in individual experience, and as such
    (1'c) are 'mappings' of that experience.

So, slightly to reformulate the whole matter again, it may be said that according to the presentation so far, constructivism affirms the following:

(A) As regards 'reality', the latter is, at least insofar as it is knowable,
    (1) the experience of an individual subject, and
    (2) a 'construction' of that subject.
(B) As regards knowledge, it
    (1) consists of concepts,
    (2) is a 'construction' or 'result of a construction',
    (3) the construct being a 'mapping' of what is 'viable'/'feasible' in the experience referred to under (A) above.

These theses must now be looked at somewhat more closely.
5. (B) may be taken first, as the issues here are, if not all straightforward, at least more so than those under (A).
5.1. (B1) need not detain, as in some sense of 'concept' it is clearly

true. But (B2) needs attention. To start with, what is meant by 'construction'? Since this idea also arises in the context of (A), to be considered further on, the general question may as well be broached here. No technical meaning has been assigned to the word in the paper being examined. Therefore there is nothing for it but to concentrate on the standard ordinary meaning. According to the *O.E.D.*, the primary meaning of the verb 'construct' is: 'To make or form by fitting the parts together; to frame, build, erect'; 'construction' as a verbal noun is the action of so doing, and, as denoting the product rather than the process, it means 'a thing constructed'. Spelling this out (with an eye to the Aristotelian 'four causes') it may be said that 'construction' involves: (1) some preexisting material ('the parts'); (2) some principle (as it were) of construction (governing the 'fitting together' of the parts); (3) some executor (of the making, forming, fitting together); (4) some end-in-view (since normally constructing is done for some purpose, as a house for shelter, ostentation, or whatever).

In the next place, what, more precisely, about knowledge is supposed to be a 'construction'? Two possibilities immediately present themselves, namely, the constituent concepts and the statemental/propositional complexes from concepts which may be affirmed or denied to the 'viable', and so on (of course, the conjunction of these two possibilities is a third possibility). Now of the first possibility it may be said that we can distinguish, even among our quotidian concepts, (a) some that we just normally come by, so to speak, 'spontaneously' formed concepts like 'red', 'loud', 'rough', even 'ugly', 'decent' and so on, and (b) others that are more or less artifically formed, like 'bachelor', 'motor car', 'immigration'; and of course the second sort increase in significance the further we move towards the heart of scientific concept-formation proper. Though the first may be said to be 'constructions' from the point of view of later scientific analysis (logic, psychology, linguistics, and so on), it is the second sort that may be more commonsensically called 'constructions' in the light of the ordinary dictionary definition, the application of which here is fairly straightforward, at least for present purposes. Coming now to the second possibility as to what may be properly called a 'construction' in the context of knowledge, namely, the statemental/propositional one, it is clear enough that it is in order to make a distinction here similar to that just made in the case of concepts. More specifically, we just 'pick up' certain ways of putting words together (e.g. 'The dog barked') while we have to learn other ways (e.g. more or less complex concatenations of clauses, rephrasings in the terms of a canonical logical notation). Again we can properly call the second 'constructions' in a quite straightforward, commonsense way, in accordance with the dictionary definition. So each possibility is open for understanding how knowledge is a construction (and hence both). So far, so good.

5.2. What about (B3)? There are at least two major questions of interpretation here. One is how 'mapping' differs from 'corresponding'. The other is how 'viable'/'feasible' differs from 'true', or, speaking more exactly, since the former seem to be clearly evidential/epistemic concepts whilst the latter is, in the way it is used in Glaserfeld's paper, anyway, an 'ontological' one, and are hence not categorially commensurable, how does 'viable'/'feasible' differ from what are frequently, indeed probably usually, taken as evidential/epistemic correlates of truth, like 'verified', 'confirmed', and the like?

5.21. As regards the first question, the paper being examined supplies no material whatever on which to base an answer. 'Mapping' is not explicitly introduced as a technical term, and its ordinary meaning involves 'correspondence', at least in the sense of a rule-governed 'projection'-relation between what maps and what is mapped. If it is suggested that a map need not 'resemble' what is mapped, in the way in which even the worst photograph of something resembles it (otherwise it would not be called a photograph) it may be rejoined that it is at least very doubtful if even the most conservative proponent of a traditional 'correspondence theory of truth' thinks that, say, 'In a closed system, entropy tends to increase with time' *resembles* something in the preceding sense. (What such a person would say more nearly about the alleged 'correspondence' relation would also be certain to be obscure or seemingly vacuous, but that is not the point here.) So it seems as though we must take the key term 'mapping' as a primitive, and people are free to attach whatever subjective meanings to it they wish.

5.22. As regards the meaning of 'viable' (and hence 'feasible', since this is linked with 'viable'), though this is not explicitly introduced either, and its ordinary meaning is too loose to function satisfactorily, just as it is, in an epistemology, there is the clue offered by the remark that it is synonymous with Piaget's 'adapted' (125), keeping in mind that it is also said that the constructivist 'orientation was ... propounded ... by Piaget as a developmentally grounded constructivist epistemology' (121). More specifically, Piaget's significance for constructivism is summarised as follows:

> ... knowledge for Piaget ... is the collection of conceptual structures that turn out to be adapted ... within the knowing subject's range of experience. ... cognitive change and *learning* take place when a scheme, instead of producing the expected result, leads to perturbation [= 'disappointment' 127 – WAS], and perturbation, in turn, leads to accommodation that establishes a new equilibrium. Learning and the knowledge it creates, thus, are explicitly instrumental ... His theory of cognition involves a two-fold instrumentalism. On the *sensory-motor* level, action schemes are instrumental in helping organisms achieve goals in their interaction with their experiential world. On the level of *reflective abstraction*, however, operative schemes are instrumental in helping organisms achieve a coherent conceptual framework that reflects the paths of acting as well as thinking which, at the organism's present point of experience, have turned out to be viable. The first instrumentality might be called 'utilitarian' ... The second, however, is strictly 'epistemic'. (125, 128, 129)

It is impossible to avoid a feeling of profound anticlimax on concluding this passage, for far from registering the radical 'shift of epistemological presuppositions' (121) promised, in essence it sets out simply some central features of a fairly standard, middle-of-the-road, more or less recent empiricist position. That this is the case is concealed, to the extent that it is (surely only to the neophyte in such matters), by the use of relatively unfamiliar terminology, and any effect produced by this can be cancelled just by a translation back into the original, using a dictionary containing such entries as: 'adapted' = 'confirmed', 'scheme' = 'theory', 'perturbation' = 'disconfirmation'/'falsification', 'accommodation' = 'theory-change/modification', and so on and so on. The page on which it is written shows itself to be a sort of palimpsest where an empiricist text has been written over by a 'constructivist' translation, and an empiricist would be justified in doing just a little rewriting himself and exclaiming, after Horace:

> Mutato nomine de me
> Fabula narratur.

So once more the examination has drawn a blank in the search for a more than sloganistic presentation of 'constructionism'. Perhaps it will do better with the theses assembled under (A) at the end of Section 4 above. Anyway, to these it now turns.

6. The theses in question pose at least two basic questions, namely, what does it mean to say, and why should we say (1) that to talk of 'reality', at least of that 'reality' which is accessible to knowledge[3], is to talk of the experiences of individual subjects, and (2) that this 'reality' (experience of the individual subject) is a 'construction' (of that subject)?

6.1. To begin with (1), it must be said to start with that constructivism is again only following a completely traditional form of empiricism in speaking of the object of knowledge as being 'experience' (or something that adds up to the same thing). Now familiarity with this habit should not breed contempt for at least the following facts. First, this 'experience' is never normally introduced as a technical term. Second, it is therefore left to be understood in at best some ordinary, everyday sense or other. Third, what is normally taken to be known is never 'experience'. If we look at the history of the word in English we find that the verb is derived from the noun, which itself originates (proximately anyway) in the Latin verb *experior*, meaning to try or test, and that the English noun preserves this sense virtually exclusively till about the last third of the eighteenth century. (Dr Johnson's *Dictionary* of 1755, for example, lists no other meanings than 'practice; frequent trial; knowledge gained by trial and practice'.) There is nothing especially pertaining to 'consciousness' here. After the time mentioned this meaning persists, but is joined by another which involves reference to a particular kind of consciousness which can in some contexts be distinguished from or even contrasted with reasoning, conscious experiment, knowledge, including as it does feeling as well as thought (as in 'aesthetic experience', 'religious experience').[4] But this is rather the exceptional case and in general it is even here a matter of the experience *of* something (e.g. religious experience has to do with God). So in general the ordinary noun 'experience' has an objective 'intention' (experience *of* something) and the verb is basically transitive (someone experiences *something*). The experience is a *means to* knowledge not the *object of* knowledge. Even where we might be inclined to say that our experience is something more or less purely subjective, as in the case of an hallucination, which someone might begin describing by saying: 'I had a strange experience', the noun is being used in a 'logically secondary' way, that is, roughly, in a manner the understanding of which depends on understanding the primary objective sense: it is of the essence of an hallucination that it is *like* 'the real thing'.

Much more could be said on this head, but I hope the thrust of the discussion is clear, namely, that insofar as someone is using 'experience' in its ordinary meaning there is nothing normally 'subjective' about it, and so some special arguments must be provided to justify a contrary view of the matter. Historically, arguments of this sort were offered by, for example, Bishop Berkeley. But the paper under examination does not offer any such arguments and does not even hint at what they might be. In view of the existing body of powerful argument against taking the object of knowledge to be 'experience', even against the intelligibility of this,[5] it is utterly inadmissible in a serious philosophical discussion not to offer grounds for rejecting these arguments if the acceptability of the position depends upon their disposal.

6.2. To pass to the second question posed at the beginning of this section, that of 'reality' = 'experience' being a 'construction', strictly speaking, if the first question is unresolved, as 6.1 has suggested it is, then *caedit quaestio*. But a few remarks may be offered anyway.

As in the case of 'experience', 'construction' is not introduced explicitly as a technical term and so once more we are by default thrown back on the ordinary meaning of the word. This ordinary meaning has been outlined at the beginning of 5.1 above. Applying this to the present case we must ask first what the materials of the construction are supposed to be. Presumably 'experiences' ('sense-impressions', 'sense-data', or whatever). But the question of the intelligibility of the latter and then the question of the justification for introducing such an idea, assuming it has been assigned a meaning, has not been addressed. Second, since both plan-of-construction and end-in-view presuppose a constructor, critical attention may be focussed on this factor in the ordinary meaning of 'construction'. Hume, in one of the most poignant passages in the philosophical literature, admitted frankly, in a note at the end of his *Treatise of Human Nature*, that he could finally make no sense, in terms of his fundamental sense-impressions, of the self to which they are referred, since he could not identify any impression of a self. More generally, if reality is a construction in subjective experience, then each constructing self must be the construction of another such self, *ad infinitum*. Or, if this consequence is to be avoided, there must be an unconstructed constructor (a *constructor sui* as it were). Then if so, why can there not be an unconstructed *object?* But this sort of metaphysical dialectic could be spun out indefinitely and should be postponed till constructivism has succeeded in assigning some intelligible meaning to the notion of reality's being the result of a subjective construction of or in experience.[6]

6.3. But is this too premature? It might be objected that since the paper says that the constructivist 'orientation was proposed by Vico at the beginning of the 18th century' (121). Vico is a so far neglected source for information about the idea of 'construction'.

6.31. The paper says that according to Vico

> epistemic agents can *know* nothing but the cognitive structures they themselves have put together.... 'To know' means *to know how to make*. ... one knows a thing only when one can tell what components it consists of. Consequently, God alone can know the *real* world, because He knows how and of what He has created it. In contrast, the human knower can know only what the human knower has constructed. (123)

This is unfortunately all that is said. Before commenting on it two things should be pointed out. First, it is not absolutely clear whether the author simply means to report Vico's views here, without necessarily endorsing them all, or whether it is also to be taken as a statement of his own views. Second, whatever light it might cast on the idea of cognitive construction in general, it does not mention anything about individual subjective experience, so that the bearing of the former on the latter must remain conjectural.

This having been said, it requires little more than a superficial reading to see that the passage presents a vertiginous array of problems of interpretation, only some of which can be touched on here.

(1) The second and third sentences seem to be the key ones: the first leads up to them and the last two are presented as consequences of them. But they do not hang together, either severally or in conjunction. Severally they are vacuous, because in the first 'to know' is explicated by a locution containing that verb essentially, and in the second 'know[s]' is explicated wholly or partly (it is not clear which) by a locution containing 'tell' essentially and this can only here mean 'know'. Looked at in conjunction, the second of the two sentences would seem to be meant as an explication of the first, but states in fact at most a necessary condition for it. In fact if 'tell' is read as something like 'identify' it plays no significant part in the satement as a whole.

(2) The first and last sentences would seem to go together, and read thus say that human knowers can know only 'cognitive structures'. Now on the face of it the latter phrase must surely mean something like 'conceptual structures', for 'real' structures, in the sense of extra-discursive ones, are surely not themselves 'cognitive', but are what 'cognitive structures' proper are used to know *about*. So it seems that human knowledge is restricted to concepts and their concatenation, whereas it has seemed so far that it is 'experience' that is known as well.

(3) Finally, the fourth sentence introduces what would seem to be an implicit contrast, namely, between '*real*' and something else, but does not even hint at what the contrast is with ('apparent' world, world of concept?) and the so far unmentioned idea of creation, which may most probably be taken to refer to that of bringing into being *ex nihilo*.

Again the analysis could be continued for some time, but there seems little prospect of seeing anything very clearly through this glass, dark as it is. So it will be better to turn to Vico himself, who, though a notoriously obscure writer, could hardly offer worse interpretative difficulties than the passage cited above.

6.32. Vico's basic theses, insofar as they bear on the present questions, may be put as follows.
1. Someone can know in the strict sense of the word (be in possession of *verum*, formulated in *scientia*) if and only if what is known is made by (*factum*) that knower, as regards both the elements and the relations that constitute it, where 'made' means brought about *ex nihilo*, 'created'.[7] (This will be called henceforth 'Vico's Principle', or generally 'VP' for short.)
2. Therefore only God can know everything, without exception; in particular, only God can know the natural world. This he does through the exercise of *intelligentia*.[8]
3. In general, human beings can at best attain *certainty* where the natural world is concerned, formulated in *conscientia*, through the exercise of *cogitatio*, about disposition of the elements of nature.[9]
3'. The qualification 'in general' in (3) alludes to the fact that Vico qualifies his thesis about the scope of human knowledge of nature in his remarks on experiment. However, his language is even less clear than usual at this point, and it is ultimately unclear what precisely he means. He says that experiment permits human beings to create (in some sense) new states of affairs in nature, and the most likely reading is that experimental 'knowledge' is supposed to lie somewhere between merely observational information and Divine knowledge proper.[10]
4. The conditions for knowledge proper, formulated in VP, are satisfied in the human domain at only two points, namely, (a) mathematics, and (b) 'the common nature of the nations', that is, matters social (including, of course, history).[11]

6.33. What light, if any, does 6.32 cast on the character of the constructivism in the paper being examined? Unfortunately, the answer can only be: at very best, precious little. It might be weakly conjectured that denial of the possibility of humans' knowing the really and truly real is related to (3) above, and hence ultimately on some version of VP in (1) and the consequence drawn in (2). As regards (3'), the paper does not mention experiment, nor is (4b) remarked upon. The most that it is possible to squeeze out is something perhaps in agreement with (4a) when it says of 'deductive inferences in logic and mathematics' that 'in Piaget's view' – and presumably according to the constructivism represented by the paper

– 'the certainty of conclusions in these areas pertains to mental operations and not to sensory-motor material' (129). So the detour through Vico has not resulted in an advance much closer to the goal of a reasonably clear presentation of constructivism, and in particular what 'construction' means in the case of 'individual subjective experience', on account of which recourse was had to Vico to start with.

7. The overall results of §§2–6, which have been devoted to an attempt to 'construct' a clearer picture of constructivism than is available in the paper under examination, have been very disappointing; it has not proved possible to clarify the doctrine to any significant extent. Indeed, in some ways it is more obscure now than at the outset, any intuitive intelligibility it may have had at the outset having evaporated along the path of a search for a more explicit understanding. So one is inclined to say of the doctrine what Falstaff said of Mistress Quickly:'Why? she's neither fish nor flesh; a man knows not where to have her'. Indeed it is like mathematics, at least in the way Russell once characterised it 'as the subject in which we never know what we are talking about, nor whether what we are saying is true' (Russell 1956, p. 1577). However, the next part of the discussion will be devoted to some critical discussion as far as possible independent of questions of presentation of the elements.

II

> What then shall we say to this?
> *Romans* 9:30

8. I propose to argue that Vico's conception of knowledge in the strict sense, as focussed in VP is, if 'knowledge' is being used here in any sense identifiably related to the ordinary range of meanings of the word, and not in some special sense which has not been made explicit, simply unintelligible. One main reason for this is that 'knowledge' in any ordinary, understandable sense of the word requires something other than the knowledge of which the latter can be said to *be* knowledge. Take, for example, Vico's example of mathematics, on which most have a better intuitive grip than the theological idea of creation *ex nihilo*. It is said that we can know this domain in the strict sense of VP because we make the truths it contains. But it does not make sense to call a pure posit an 'object' of 'knowledge': the person laying down the stipulations is doing just that, and this is no more 'knowing' than conferring a name on an infant is 'knowing'. Of course, there can be knowledge *that* such a posit has been made – or that such and such a name has been conferred – but this is obviously a quite different matter. Again, the posits having been made, it does not make sense to speak of 'inferring' from it (an axiom for example) unless the inferring is constrained by some rules, just as it does not make sense to say that someone is 'playing chess' if that person just makes his moves as he pleases. Of course, the rules can be changed, but they do not become something we can be said to 'know' until the change has been made, and we cannot be said to 'infer' till the change has been adopted, however temporarily. Again, to take the conclusions derived by inferring from initial posits, these are no more 'true' (rather than correctly describable as, say: 'derived in accordance with the rules of inference and premises of the system used') than a winning end-game in chess is 'true'. We can talk of truth value once the system has been applied to some extra-mathematical subject matter, as in 'applied geometry'; but then human beings do not 'make' that subject matter, in its ultimate constituents anyway, and so, according to VP we cannot be said to have access to 'the true' and hence to knowledge. Similarly, in the case of 'the common nature of the nations', we can *decide* what is to be called 'just', say, and draw conclusions, but this does not vouchsafe any privileged cognitive access to that act (say) which is (correctly) called 'just', with respect to, for example, its actual consequences, which are not generated by that act qua 'just' any more than we can infer something about a man who is correctly called 'married' other than that he stands in a certain relation to a 'wife'.[12] Examples from other areas could be multiplied,[13] but they would only reinforce the contention that it is nonsense to talk of knowledge in the strict sense entailing creation of what is known or of an identity between knower and known. Rather, it does away with the idea of knowledge altogether, just as (to use Kant's image) though birds may well fly better the less the air-resistance, they could not fly at all if there were no air at all.

9. It may be worthwhile to consider what might be described as a more 'relaxed' version of VP, one where the actual 'creation' of the elements and relations are not in question. In this version knowing and making are said to be necessary and sufficient conditions for one another. Is this plausible?

9.1. Consider the matter from the side of knowing. First, it is obviously not sufficient, otherwise manufacturing would be cheaper than it is. But, second, is it at least a necessary condition? To make the discussion non-trivial, suppose it is assumed that a necessary condition for 'making' here is some sort of conscious intention, planning, intentionally used procedure, so excluding whittling (and much modern 'art'). Then suppose that a normally competent chemist sets out to synthesise a certain compound, but instead ends up with another (perhaps being thereby responsible for initiating a breakthrough in both knowledge and techniques of synthesis). Would it not be correct to say that he 'made' the substance he actually produced? To point the question, bringing it closer to Vico's concerns, consider a similar case in the historical field, one from the vast repertory of 'unintended consequences'. Vico is one of a number of thinkers about history who have devoted much attention to this theme, belonging to that line of thinkers who, rather than looking at the matter in secular fashion (e.g. Mandeville, Turgot, Marx) used the idea (like Bossuet and Hegel) to reconcile genuine human agency and Divine Providence (Vico 1961, §1108, pp. 382f). But then, if the human agency is genuine, the act was not sufficient for knowledge of it and hence the latter not necessary for the former. (But if human agency is not, then the reconciling project founders.)

9.2. Let us look at the question from the point of view of the making. First, making is surely not a sufficient condition for knowing. For instance, Galileo made telescopes that were good enough to permit him to make path-breaking astronomical discoveries at a time when next to nothing of the optics of the instrument was known, and there was a similar situation with regard to the early modern steam engines *vis-à-vis* the principles of thermodynamics. Second, is making a necessary condition for knowing? It all depends, in particular on how closely associated the 'making' and the 'knowing' have to be for the one to count as a necessary condition of the other. For even if the two may in general be related, the relation may hold only by means of a very intricate and extended chain of steps. For example, there is no doubt that Euclidean geometry was ultimately derived from practical measurements of actual lengths, areas, volumes, but the path between the two was certainly a long one, and may never be at all fully reconstructed.[14]

10. Instead of looking for and at basic concepts and premises, the examination may be directed at consequences. If there is at least one of these that is in some degree intelligible, central, and also vulnerable to criticism, then, assuming that the argument, whatever it is, is valid, there can be some rational confidence that something is wrong further up the inferential line.

In looking for a starting-point of this sort it is hardly possible to do better than consider a spot that von Glasersfeld himself considers to be at least *prima facie* in need of defence, or at least of elucidatory amplification, namely, the question of intersubjectivity.

10.1. He writes:

> To make the Piagetian [= in effect, here, constructivist – WAS] definition of knowledge plausible, one must immediately take into account ... that a human subject's experience always includes the social interaction with other cognizing subjects.... But introducing the notion of social interaction raises a problem for constructivists. If what a cognizing subject knows cannot be anything but what that subject has constructed, it is clear that, from the constructivist perspective, the *others* with whom the subject may interact socially cannot be posited as an ontological given ... [constructivists] want to avoid assuming any cognitive structures or categories as innate. Hence there is the need to hypothesize a model of the conceptual genesis of 'others'. (126, 129)

10.2. Before the model in question is outlined and examined it may be noted that there is at least one not inconsiderable oddity in this passage, namely, the fact that it says that 'a human subject's experience *always* [emphasis added – WAS] includes the social interaction with other cognizing subjects', whilst it also says that 'what a cognizing subject knows cannot be anything but what that subject has constructed', so that, since 'social interaction' presumably involves knowledge, each 'human subject' must construct 'other cognizing subjects', the consequence is (a) that the subject always experiences/knows other subjects and also (b) that there is a time (namely, that before the construction of others is completed) when the subject does *not* experience/know others. If this is not a self-contradiction then it may serve as an example of such until a better one presents itself. In the fairy story we read that, though in an impossible situation, 'with one bound Jack was free'. We must await Jack-the-constructivist's self-emancipatory leap here.

10.3. Let us turn to the 'model' of how the individual subject is supposed to 'construct' others. Well, once upon a time in every very small child's life it begins to construct concepts of objects – or rather of 'objects':

> On the sensory-motor level, the schemes a developing child builds up and manages to keep viable will come to involve a large variety of 'objects'. There will be cups and spoons, . . . rag dolls and teddy bears – all seen, manipulated, and familiar as components of diverse action schemes. (129)

But, furthermore,

> there may also be kittens and perhaps a dog. Though the child may at first approach these items with action schemes that assimilate them to dolls or teddy bears, their unexpected reactions will quickly cause novel kinds of perturbation and inevitable accommodations. The most momentous of these accommodations can be roughly characterized by saying that the child will come to ascribe to these somewhat unruly entities certain properties that radically differentiate them from the other familiar objects. Among these properties will be the ability to move on their own, the ability to see and hear, and eventually also the ability to feel pain. The ascription of these properties arises simply because, without them, the child's interactions with kittens and dogs cannot be turned into even moderately reliable schemes. (129f)

10.4. Now it is difficult to know where to start in criticising this, such is the *embarras de choix*. At any rate, many of the points will have to be made quite summarily, though reference will be made to places where the argumentative underpinning is presented more fully.

10.41. If the concept of object, or of 'object', is to be constructed then there must be some pre-'object' concept or concepts *from* which it is constructed. Presumably this conceptual material refers to individual sense contents: 'sense impressions', 'sense data' have been among the terms traditionally used for the latter. But there are powerful arguments, stemming more recently mainly from Wittgenstein, to the conclusion that there is no sensation language (as it may be called) independent of and prior to object language, but that insofar as the latter exists at all it is 'parasitic upon' the latter. Much earlier, Kant argued, against Hume in particular, that 'experience' is inherently intersubjective or object-related.[15]

10.42. Essentially the same point can be looked at from a slightly different direction. If there were to be pre-object experience then it would presumably have to be the experience of some subject, some self. But, as already mentioned, Hume himself frankly admitted that he could not accommodate such a self within his system in which 'impressions' are fundamental. Kant (on whom von Glasersfeld calls in another, related regard, 130) argued that empirical subjects necessarily presupposed empirical objectivity.[16]

10.43. Again, it has been strongly argued by many philosophers[17] that, far from its being the case that the individual subject ('I', 'ego') 'constructs' or 'infers' other such, it is rather the *community* of subjects/others that constitute individual subjects, or, better, that self and other are correlative. But there are no others at this stage of the story which is, in fact, being told in order to try to explain how the idea of others comes to be.

10.44. Apart from what has been said so far, the whole constructivist story so far is still circular, since the story of the alleged process of construction assumes precisely what is supposed to be being constructed. For instance, how could the child 'assimilate' kittens and dogs to rag dolls and teddy bears unless it already had a concept of items of the first sort as being different? (Set aside here is the question of the plausibility of ascribing such a complex, sophisticated process to a small child.) And why should it conceive kittens and dogs significantly differently from rag dolls and teddy bears in the respects listed? If we are dealing purely with what belongs to the level of external phenomena (and consistency with the program of constitution makes this mandatory at this stage) then nothing about what we call 'kittens' and 'dogs' requires the postulation of self-movement, ability to see and hear and feel pain. All this presupposes what has been traditionally called an 'analogy argument', which not only has been completely discredited,[18] but is one which is merely risible to imagine a child able to construct.

10.5. Finally, there is said to be 'a very similar development' (that is, to the one set out in 10.3 above) to

> the child's construction of schemes that involve still more complex items in her experiential environment, namely the human individuals who, to a much greater extent than other recurrent items of experience, make interaction unavoidable.... Here again, in order to develop relatively reliable schemes, the child must impute certain capabilities to the objects of interaction. But now these ascriptions comprise not only perceptual but also cognitive capabilities, and soon these formidable 'others' will be seen as intending, making plans, and being both very and not at all predictable in some respects. Indeed, out of the manifold of these frequent but nevertheless special interactions, there eventually emerges the way the developing individual will think both of 'others' and of him- or herself. (130)

Nothing essential needs to be added to the considerations set out in 10.44 above, for, *mutatis mutandis*, they apply here too. If the gap between what we have available for 'construction' and what is supposed to be constructed is obvious enough there, then here it becomes a 'dark unbottom'd infinite abyss'.

11. It is necessary now to look at the constructivist account of language acquisition both for its own sake and because it bears very closely on the subject matter of §10. It is written:

> From the constructivist point of view ... language users must individually *construct* the meaning of words, phrases, sentences, and texts. Needless to say, this semantic construc-

tion does not always have to start from scratch. Once a certain amount of vocabulary and combinatorial rules ('syntax') have [sic] been built up in interaction with speakers of the particular language, these patterns can be used to lead a learner to form novel combinations and, thus, novel conceptual compounds. But the basic elements out of which an individual's conceptual structures are composed and the relations by means of which they are held together... must be abstracted from individual experience; and their interpersonal fit, which makes possible what we call communication, can arise only in the course of protracted interaction, through mutual orientation and adaptation... (132)

12. The following are some critical comments on this passage.

12.1. There seem to be at least two quite different accounts of language acquisition here. (a) According to one language users must *individually* 'construct' meanings. Now since there is a trivial sense in which this is true (they must individually acquire language just as they must individually learn to knot a tie or drive a car, since only 'I' can do anything that is 'my' doing) it may be worth remarking that this is clearly meant in a nontrivial sense. That this is so emerges already from the statement that this construction 'does not always have to start from scratch'. For if it does not 'always' have to do so, then it may at least sometimes do so. So in the basic sense language acquisition is individual in the stronger sense of arising 'from scratch', or *ex nihilo*, as we might say, remembering the shadow of Vico. Nothing innate can be assumed. (b) But then it is said that an initial fund of semantics and syntax is 'built up in interaction with speakers of the particular language'. But perhaps what is meant is that this occurs in those cases when the individual learner does not 'start from scratch'. Anyway, let us consider both possibilities.

12.2. To take (b) first, something very strange immediately emerges once it is recalled that the passage also says that the 'interpersonal fit' between individually constructed languages 'which makes possible... communication' depends upon 'protracted interaction... mutual orientation and adaptation'. Now it seems to follow from this that the initial 'interaction with speakers of the particular language' which is a presupposition of the construction of a basic semantics and syntax must occur *before* 'communication', or, still more explicitly, that learning language with the help of others occurs before communication with others. Who can make anything of this?

12.3. According to account (a), semantic and syntactic 'basic elements' are (i) 'abstracted' from (ii) 'individual experience'. Now there are many problems here, major ones clustering about the aspects of (1) abstraction, (2) individuality and (3) experience.

12.31. The received discussions of 'abstractive' theories of concept formation are made up of a number of threads, many of which are independent of one another, as, for example, those concerned with the abstractive theory specifically of generic concepts and with psychologistic theories. One perhaps most relevant here, and going to the heart of the matter, is the following: abstraction theories in general are essentially circular because the alledged process of abstraction already *presupposes* the concept which is supposed to be formed as a *result* of that process. For example, if I am supposed to learn the meaning of 'red' by 'abstracting' the common property 'redness' from various items that are red, this assumes that I can already form a class of 'red' things from which the abstraction can be made. But this clearly assumes that I am competent to distinguish red things from others. Or, if we imagine the class already formed it assumes that I can pick out *that* common property rather some other, since any group of objects at all possesses a *number* of common properties. The situation is not changed in principle if, going beyond the individualist context, we imagine that someone exhibits an instance of a property the word for which it is desired I should learn (points to a *red* object, makes a *loud* noise, and so on): the process of so-called 'ostensive definition'. But here too, since anything always instantiates different concepts at the same time, I must already have at least a rudimentary command of the concept intended if I am to realise what is meant.[19]

12.32. The last consideration already introduced others into the language-learning process, and it is at least one strand of Wittgenstein's much-discussed analysis of the idea of a 'private language' that the latter, interpreted strictly is impossible (certainly as a 'first' or 'basic' language, as it were, as distinct from one which I might make up for a special purpose – for example, writing a diary in – but which would depend on a more fundamental one). This is (roughly) because to have a language in any genuine sense of the word requires that I achieve consistent reference, for the most part at least: a system of markers (written, auditory, etc.) that did not 'mean' much the same thing for most of the time not only could not be used to communicate with others, but could not be used by me to communicate with myself. So a community of language users is necessary to provide the possibility of intersubjective checks on consistency of reference.[20]

12.33. Furthermore, a second strand in Wittgenstein's private language argument connects with the third point of focus of problems identified at the beginning of 12.3, namely, the constructivist idea that basic concept formation is not only abstractive and individualist, but also takes place in the field of experience. For Wittgenstein argues, in effect, that consistency of reference presupposes not only a fairly stable world to which reference may be made, but also a commonly accessible, public world to which recourse may be had for purposes of checking consistency of reference.[21]

12.4. One last point may be made about this question of the relation of language and the experience of a unified and differentiated world. This is that there is excellent scientific material pointing to the conclusion that a fully coherent experience of such a world depends on at least some mastery of language,[22] and that hence any account that separates such symbolic processes and intersubjectivity must be seriously astray.

13. At this point we may pause to notice that there is a common feature in the paralogisms exhibited so far in this part of the examination, namely,

a pervasive *inversion* of *terminus a quo* and *terminus ad quem*. These inversions are mainly the following:
1. of sensation and object;
2. of ego and object,
3. of ego and others;
4. of abstraction and possession of concepts;
5. of individual language-acquisition and intersubjectivity.

All this might cause us to be reminded of the slogan quoted with approval from Piaget: 'intelligence organizes the world by organizing itself' (136). If we must have a slogan, then let it be rather: 'intelligence organises itself by organising the world'.

14. Finally, coming right back to the beginning again, we saw (3.2) that, in the paper being examined, two notions of objectivity are distinguished, namely, one relating to the world and the other to intersubjectivity, and that constructivism in effect proposes to drop the first whilst keeping the second. The burden of the examination so far is that this is impossible, and that to take the first step, in the direction of 'subjective reality', is to enter upon the path to 'the undiscover'd country from whose bourn/No traveller returns...'

III

> Confusion now hath made his masterpiece!
> *Macbeth*, II.iii:72

15. For crime stories to be really satisfying it is necessary that they not only reveal the villain, and the course of the process of inquiry that led to that discovery, but also how the villain came to commit the crime in the first place. Similarly, a critical examination of an intellectual position is the more satisfactory to the extent that it not only reveals errors but plausibly shows how they came to be made, not, moreover, in merely individual terms (which are generally fairly uninteresting from the point of view of the history of ideas) but in more objective ones. So, having now concluded the present critical examination in the first mode, I want to turn to the second. In the nature of the case this cannot be as compelling as the first sort can be, at its best anyway, but may be at least suggestive and even have some independent value.

16. If to start with we look once more at 'Vico's Principle' ('the true' and 'what is made' are interchangeable) probably the first thing that will strike most, after its meaning and implications have been spelled out, is what an extraordinary one it is. For it is an epistemological criterion that allows as genuine knowledge of the natural world only God's, that, in the end, bases knowledge of socio-historical matters on revelation of the character of Divine Providence, and allows to human beings by way of knowledge in the strict sense only mathematics.

Now students of philosophy are used to its practitioners' saying extraordinary things and may often be grateful if what is said is just (perhaps even with the help of a sympathetic reading) intelligible. Of the many different sort of responses to philosophers' utterances, one extreme one is to regard those making the statements as just more or less harmless lunatics, whilst another, at the opposite extreme, is to learn not to notice any more how very strange the doctrines are, perhaps even to the extent of eventually saying things of a similar sort. Now an approach lying somewhere between both, to be recommended especially when the philosopher making such statements seems to be otherwise reasonably balanced, and perhaps even often makes penetrating remarks, is to follow that 'rule of charity' that we would normally follow if someone we knew to be intelligent and level-headed were to say things we thought very curious indeed, or to act in a way we thought inappropriate in the circumstances, that is, to ask whether we had really understood the *point* of his words or behaviour, and whether, this having been grasped, his words or behaviour might not now seem quite intelligible to us, and not merely perverse, even if that rationale might still seem so, or at least unacceptable. In the case of a philosopher we should often regard some or all of his doctrines as determined by a 'hidden agenda', as answers not to overt questions but to covert ones of the sort: What answers to the former sort will produce cognitive effects which advance the cause of a predetermined position?[23]

In the case of Vico, once we ask *this* sort of question a plausible answer is fairly obvious. For VP is tailored, as it were, to the end of securing knowledge of the real world, natural and social, for Christian belief, leaving mathematics to human beings, since, as Hobbes said of geometry, 'men care not, in that subject, what be truth, as a thing that crosses no man's ambition, profit or lust . . . not . . . contrary to any man's right of dominion, or to interest of men that have dominion' (Hobbes 1651, p. 91). And, looking at the matter from the point of view of Vico as an individual, the result agrees perfectly with everything we know of his basic theological orientation.[24]

But this has still not got to the heart of the matter, for if we carry our 'sifting humour' further we must ask why it was necessary to try to secure Christian belief just at this time. Again, the question almost answers itself once it is posed. It is indeed given very explicitly in a work of similar orientation, though containing different doctrines, published in the same year as Vico's *De antiquissima Italorum sapientia* (1710), namely, Bishop Berkeley's *A Treatise Concerning the Principles of Human Knowledge*, which bears the subtitle: *Wherein the Chief Causes of Error and Difficulty in the Sciences, With the Grounds of Scepticism, Atheism, and Irreligion Are Inquired Into*, or still more succinctly in the same author's *Three Dialogues Between Hylas and Philonous*, published three years later, which announces on the title-page that the work is *In Opposition to Sceptics and Atheists*. In short, Vico's epistemology makes sense if seen as a battlefront in the bitter struggle of Christianity against the 'New Sciences', or, at least against what many took to be philosophical consequences of the latter at best simply unwelcome to Christianity and at worst inconsistent with it.[25]

17. Now it might be objected to all this (a) that Thomas Hobbes held a view essentially similar to VP, (b) that Hobbes's views were not inspired by theological considerations, that he was in fact a leading philosopher of the 'New Sciences', and (c) that Vico knew of Hobbes's work and even remarked upon its affiliations with its own.

With regard to this, (b) is certainly incontestable. Of (c) it is true that Vico knew of Hobbes's writings, but what he remarks upon concerning the nature of their common ground does not include VP, and the criticism he makes of him concerns precisely Hobbes's lack of a theological basis for his account of the history of society.[26] All this by itself would make necessary a questioning of the truth of (a). To do this it is indispensable to have at least a brief outline of Hobbes's thought on the relevant issues before us. This may be set out as follows.[27]

1. Science (*scientia*) is concerned with the truth of general propositions.
2. General propositions are ones about 'consequences', that is (so the point may be glossed), general propositions are conditionals.
3. When it is a matter of the truth of a fact we speak simply of knowledge (*cognitio*).
4. The content of science (that which *scimus*) is 'knowledge from causes, or in other words (*sive*) derived from a generation of the subject-matter (*subjecti*) by means of a correct (*rectam*) argument'.
5. The means by which we know in the scientific sense (*scimus*) to the greatest degree possible that a theorem is true is knowledge derived by means of a legitimate (*legitimam*) argument from experience of effects.
6. Both derivations are wont to be called 'demonstrations', but it is better to use this term for the first (5) because it is better, wherever possible, to use causes that are present rather than ones which are irrevocably in the past.
7. Therefore, science *par excellence* is that available to humans *a priori* by virtue of the fact that the generation in question depends on their own will (*arbitrio*).
8. Geometry is a body of items that may be demonstrated (*demonstrabilia*), for it treat of figures that we ourselves create (*creamus*).
9. 'On the contrary, since the causes of natural things lie not within our power, but in that of the divine will, and since the greatest part of them . . . is invisible, we cannot deduce their properties from causes'.[28] Nevertheless, 'from the properties of them that we do see, we can, by deducing consequences, go so far as demonstrating that this or that may be their causes. This sort of demonstration is called *a posteriori*,

and the science itself, physics'. Thus genuine (*vera*) physics, which depends upon (*innitur*) geometry should be included among branches of (mixed) mathematics (*mathematicas mixtas*)', for it is usual to call 'mathematics' those sciences that are taught 'not by practical use (*usu*) and experiencing but by teachers and by rules', and thus 'pure' mathematics is that which treats quantities *in abstracto*, there being no need of knowledge of a particular subject-matter (*subjecti*), whereas 'mixed' mathematics are those where such knowledge is required.

10. Politics and ethics, that is the science of just/unjust, equitable/inequitable, can be demonstrated *a priori* because the principles by virtue of which what these are are known, that is, their causes, namely, laws and agreements, we ourselves make (*fecimus*).

18. If this account and Vico's are now compared it will be seen that they are fundamentally different.[29] Though Hobbes speaks of 'generation', and though this is absolutely central to his views here, it must be emphasised that this is not at all the same as Vico's 'making', even if the two terms may seem to be at least very similar in meaning and even overlap in this respect at the point now in question. For Hobbes's 'generation' is *intra*discursive, its being a matter of the development of a subject-matter from – ultimately – definitions, whilst Vico's 'making' is, at least as regards nature and the socio-historical domain, *extra*discursive. The character of Hobbes's view is probably somewhat obscured, for modern readers anyway, by the fact that he uses 'cause' in such a way that it covers both what we might call today 'reason' and also real cause.[30] (Perhaps 'explanation' in current usage has some of this ambiguity.) Still, 'causes' here are *internal* to discourse,[31] even if they in some sense correspond to real causes. Mathematics on the one hand, and politics and ethics on the other, belong together for Hobbes not because we generate or make the subject matter of both (the ideal objects of the one and the real objects of the other), but because we generate or make the ideal objects of both, in such a way that there is no gap between nominal and real essences, as it were, here, whereas this is not the case in physics. Hobbes does not deny that there is a science, in his strict sense, of physics, only that there is no *a priori* science of physics. From knowledge of effects physics can strictly demonstrate a disjunction of possible causes, but cannot pick out *a priori* which of these disjuncts is, in such and such a case or type of case the actual one. Though he does not say explicitly that this task of determining actual causes from a set of possible ones falls in the domain of empirical inquiry, that is presumably his view.[32]

19. Furthermore, it should now be clear that Hobbes's view here is an inheritance from Galileo, rightly regarded as the founder of modern natural science as regards its method (insofar as there is a single 'founder') and the central figure therein substantively.[33] Hobbes takes his central ideas in this regard from Galileo but in return clarifies and generalises them, making them available for philosophical thought in general. To sharpen this claim let us recall the principal features of Galileo's approach.

This may be best done by bringing them into relief against the Aristotelian approach he combatted.[34]

According to the latter, (1) the subject matter of science is ultimately the world as ordinarily perceived and its task the recording of what is the case there 'always or for the most part' (*Met.* 1026b28–1027a28). (2) The task of this recording is taken on by generic (class-) concepts, abstracted from the world as ordinarily experienced and underpinned by 'formal' (in general equivalent to final) causes.

For Galileo, (1) the everyday world simply does not immediately exhibit significant regularities, far less necessary coexistences and sequences. Rather, it has to be interfered with instrumentally (experimentally) so as to produce systems as closed as far as possible against certain causal factors: only in such relatively isolated systems is there the possibility of observing invariant behaviour. But the indispensable guide for this practical interference with the actual, which makes the basis of physics not so much perceptual as experimental experience, is (2) a set of special concepts, which must be formed, in the order of knowledge, in advance of experiment, concepts that refer to purely hypothetical situations and thus cannot be 'abstracted' from experience of the ordinary world. They are not generic but rather relational concepts expressing not 'formal' or 'final' causes but *conditions*, realised in 'efficient' causes, the totality of which constitute the state of affairs in question. The instances of such concepts do not simply stand side by side, as it were, as in the case of generic concepts, but are related to one another by strict rules of dependence. Experiment on the one hand, and concept and theory formation on the other hand, are reciprocally related.[35] The latter guides the former, but the former is the ultimate judge as to which of the in principle competing reconstructions in theory of the actual from different sorts of analytical concepts and principles (which 'composition' from a preceding 'resolution') is the more adequate. In sum, Galilean methodology is constructivist in at least the following ways: (1) it involves the practical (re)construction of the actual world in experiment; (2) it involves a construction of concepts in the sense that it cannot make do with ones that are formed in a more or less spontaneous way in everyday life but must tailor ones suited to the representation of situations which do not spontaneously occur in quotidian experience, but have to be experimentally produced; and (3) still within the domain of concepts, these govern fields the particular items in which can be 'constructed' one from another by the use of rules that generate series. Indeed, bringing all three points together it may be said that the ultimate point of the enterprise of physics is the transformation of a merely 'empirical' manifold of simply coexisting or successive elements into a 'constructive' manifold of elements that are related by strict functional rules.[36]

Though Hobbes does not take up all the features of Galileo's methodology, sufficient essential points are there to make the genealogy unmistakeable: the province of science defined as the field of 'consequences', that is conditionals, rather than categorical statements about what is actually the case,[37] the consequent distinction between the realms of the intra- and extra-discursive, the conception of theory as providing a repertory of possible causes of given effects, the rejection of 'formal'/'final' causation (Hobbes 1656, Pt II, Ch. X, §7).

20. It only remains to draw the conclusion, foreshadowed at the beginning of §17 above, that 'Vico's Principle' and Hobbes's methodology are quite different in principle and confusion between them is at least partly the result of an attention to words rather than to what they mean.[39] They are not only systematically (conceptually, theoretically) quite different but (not unconnected with this of course) belong to quite different historical lineages, Vico's to traditional Catholic theology, Hobbes's to early modern science.[40] (That the personal affiliations of each philosopher were correspondingly different is also true.) Each belonged to a different part of the contemporary field of ideas, Hobbes to the intellectually revolutionary spirit of the new sciences, Vico to the conservative front against the latter.[41]

21. Finally, let us return to the doctrine called 'constructivism' which was the main subject of examination in the first and second parts of this paper. We can ask whether the considerations of the third part have provided any resources for suggesting a diagnosis or aetiology of the problems it has been shown to face, even for sketching a proposal as to the way in which, systematically speaking anyway, it came to be 'constructed'. This must be an enterprise with conclusions that are tentative at best, but for what it is worth the following is suggested.

According to traditional empiricism,

(1)   the direct object of knowledge is given to the individual subject in experience.

Now Kantians and others had long insisted that this involved what was called in the mid-fifties (Sellars 1956) a 'myth of the given', and a great many otherwise empiricist philosophers began scrambling belatedly aboard the new bandwagon that had inscribed on its side the following words:

(2)   The object of knowledge is not 'given' directly to the individual subject in experience but only via 'theoretical constructs'.

At this point it is only necessary to confuse the object of knowledge, that is, the experience of the individual subject, with the means necessary for knowing about this object, that is, constructs, namely objects of 'construction', to arrive at the following position:

(3)   The experience of the individual subject is a 'construct', a result of 'construction'.

So, by a sort of intellectual prestidigitation there results from (1), which can be assigned some sort of sense, and (2), which is intelligible enough (once explained, anyway), (3), which is not comprehensible.[42] But that words which are individually meaningful can be put together according to legitimate syntactical rules to obtain not just sentences that are obviously unintelligible but also ones which, if not obviously such, can be shown to be so, should come as no surprise to students of the history of ideas.

22. In conclusion, I suggest that the whole preceding examination points to the following outcomes. Firstly, much of the doctrine called 'constructivism', as presented in this paper (especially in the first and second parts) is simply unintelligible. Secondly, to the extent that it is intelligible enough to provide some foothold for understanding and criticism it is simply confused. Thirdly, there is a complete absence of any argument for whatever positions can be made out. In all these respects certain words and combinations of words are repeated like *mantras*, and while this procedure may well eventually produce in some what chanting is often designed to do, namely, produce a certain feeling of enlightenment without the tiresome business of intellectual effort, this feeling nearly always disappears with the immersion of the head in the cold water of critical interrogation. Fourthly, the key problem of intersubjectivity is not successfully addressed. In general, far from being what it is claimed to be, namely, the New Age in philosophy of science, an even slightly perceptive ear can detect the familiar voice of a really quite primitive, traditional subjectivistic empiricism with some overtones of diverse provenance like Piaget and Kuhn.

It is possible that this gallimaufry might somehow inspire sounder work by others – perhaps stranger things have happened in the history of ideas – maybe by virtue of a vague resonance suggesting sounder ideas about 'construction' and so forth, but it is more likely to sidetrack thought and lead it to dead ends through its obstructing the posing of the right questions.

It may be claimed that, even if the presentation of constructivism examined here is indeed imperfect in certain ways, still there are others that do not have these failings. To that the only reply can be the one given by the people in the Greek fable to the man who boasted how far he had once jumped on the distant island of Rhodes: *Hic Rhodus, hic salta!*

NOTES

1. It may be worth noting, in order to avoid any misunderstanding, that Dr von Glasersfeld's 'constructivism' does not have anything to do, systematically anyway, with the doctrines covered by the same name stemming from what used more commonly to be called the 'Erlangen school', and also often called 'protophysics', all these largely deriving from the work of Hugo Dingler. For a brief survey of the latter's work and a comprehensive bibliography see Dingler 1987, and on the former, e.g. Böhme 1976 and Butts & Brown 1989.

2. See Pseudo-Dionysius, *Divine Names* and *Mystical Theology*, and John Scotus Eriugena, *De Divisione Naturae*, especially 1, 13 & 14.
3. There is a problem of interpretation alluded to here which must be left simply noted as such. The paper speaks of 'the thinking organism's cognitive isolation from "reality"' (121), and this must surely mean that 'the thinking organism' cannot know something about the world in general, for you can hardly be said to be 'isolated' from something that does not exist (though of course you can believe falsely that you are isolated from such). This is a form of scepticism. But the paper rejects scepticism (135), urging that constructivism evades this doctrine, and unless it does so by the trivialising move of simply defining it away by a stipulation regarding the meaning of the word 'knowledge' it is hard to see how to reconcile the two assertions.
4. For the lexicographical facts here see the *OED* and Williams 1983 s.v. 'empiricism' and 'experience'. (Dr Johnson's Dictionary, referred to above, has no pagination.)
5. Denial, or, at the very least, a heavy qualification on such intelligibility is in a sense the theme of an important part of Wittgenstein's later philosophy. (Since Wittgenstein's writings very seldom contain neat, definitive formulations of his positions and arguments, which have to be 'constructed' from different passages, it is more convenient, in a paper like the present, which is certainly not in any sense on Wittgenstein, but does refer to him at a number of places, to call upon sound secondary literature. For the present point see, e.g., Pears 1988, Ch. 11.) For an excellent recent refutation of Berkeley's arguments (if they can be called such) see Stove 1991, esp. pp. 139ff.
6. It may be noted here, in case Berkeleyan ideas were to be in the background in this area, that it is a vulgar misunderstanding of the Bishop's thought to ascribe to him the idea that individual subjects 'generate' or 'construct' their experience. Certainly their perceptions exist 'in' their minds (whatever that means) but their cause is 'the *immediate* hand of an *almighty* agent' (*Principles of Human Knowledge*, §CLI – and many other places in the same work).
7. Vico 1988, pp. 45, 46, 47, 56, 59, 64, etc.
8. On *scientia* vs *conscientia* and *intelligentia* vs *cogitatio* (see the following (2) and (3)) see Vico 1988, pp. 46f, 55.
9. Vico 1988, pp.48ff.
10. Vico 1988, pp. 52, 60, and cf. p. 97: 'as God is nature's artificer, so man is the god of artifacts'. (It is not easy to understand Vico's position on experiment, for here human beings surely only rearrange – ultimately – parts of the nature they do not make. If 'we cannot prove physical facts from causes, because the elements of natural things are external to us' as Vico says, 1988, p. 65, how is this changed in the experimental situation?)
11. On (a), Vico 1988, p. 65, and on (b) Vico 1961, §§331, 349, pp. 53f, 62f.
12. For some sound critical voices opposed to the recent choruses of praise for Vico, see Zagorin 1984 and Gaukroger 1986.
13. For instance, it used to be quite widely held that we could have incorrigible knowledge formulated in first person, present tense statements 'about' 'mental' conditions like pain, because – so it was often alleged – there is here no distinction between knower and known. But this is incoherent. For if 'I am in pain at t' is necessary and sufficient for 'I know that (I am in pain at t)' then the part of the second sentence in brackets can be replaced by the whole of that second sentence, and on *ad infinitum*.
14. On the 'prehistory of theory' in general see the very interesting and instructive Blumenberg 1987.
15. See Pears 1988, esp. Ch. 11. Also Austin 1962. (Kant's doctrine here has been widely misunderstood, to some extent because of the distinction he introduced in his *Prolegomena* (§18) between 'judgments and perception' and 'judgments of experience', obviously in the interests of popularisation, but in fact only muddying the waters, his presentation's being commonly misunderstood to mean that the first somehow precede the second.)
16. See, for example, 'the Refutation of Idealism' in his *Critique of Pure Reason*, B274ff.

17. See, for example, Hegel, *Phenomenology of Spirit*, Ch. IVA and Mead 1934.
18. See, for example, Malcolm 1958.
19. The 'abstraction' theory of concept formation goes back at least to Aristotle – see here the exhaustive presentation in von Fritz 1964 – and has been a standard one since. For a criticism particularly of psychologistic versions see Husserl 1970, pp. 337ff. The general circularity objection might be traced back at least as far as Frege, *The Foundations of Arithmetic*, §23; certainly it is developed in full in Cassirer 1923, Ch. I. See also Wittgenstein 1958, Part I, esp. 27ff.
20. See Pears 1988, Ch. 13–15. (There is a passage in the paper on pp. 130f – beginning 'that we impute the cognitive capabilities' and ending 'reality we have constructed' which might be read as a dim apprehension of Wittgenstein's point, but it is really hard to tell.) On the general question of the social context of thought and language see Vygotsky 1962 and Wertsch 1985.
21. See note 20 above.
22. See Cassirer 1959, Pt II, Ch.VI and 1985.
23. A methodology of 'question and answer' was presented by Collingwood in his 1939, Ch. V for use in historical studies in general. It jibes perfectly with something he could not have had in mind, namely, Althusser's theory of ideology. See, e.g., Althusser 1969, pp. 67ff and Althusser and Balibar 1970, pp. 52ff.
24. One illustrative passage must here stand for many. St Augustine writes: '... in illius [*sc.* Dei – WAS] naturae simplicitate mirabili non est aliud sapere, aliud esse; sed quod est sapere, hoc et esse.' (*De Trin.* XV, 13, 22). Cf. Aquinas, *S.T.* I, 14, 8. Berlin 1976, p. 117 says: 'it seems clear that the *verum/factum* doctrine is mediaeval and Christian and, by Vico's time, a theological commonplace. Also Löwith 1968, 8ff. Nicolas of Cusa already has the analogy between God's creation of the world and the human creation of mathematics. See Cassirer 1922, I, 38f, 45ff.
25. There is obviously no question of going into detail here. For some of the relevant information see, for example, Jacob 1988, Lefèvre, esp. pp. 54ff.
26. Vico simply says (1961, §179, p. 28) that Hobbes, like he himself, attempted 'the study of man in the whole society of the human race'. For the criticism see *loc. cit.*
27. Hobbes's thought here is probably most often presented in terms of a passage at the beginning of his 1656a (pp. 183f). But this is too condensed to be intelligible without a good deal of commentary and I have chosen instead to use his 1658, Ch. 10, §§4, 5, which is brief yet relatively self-contained and is, moreover, his last presentation. (The work seems not to be have been translated into English, so the slight paraphrase and translation is my own.) See also Hobbes 1656, esp. Chs I & X, and also parts of Hobbes 1660.
28. I am not quite clear what exactly Hobbes means here, and, more specifically, whether he means that not we but God creates natural causes or that only God has power over them; the coupling of this reference to God with the point of their being too small for us to see (recalling Locke, e.g. *Essay*, IV.iii.25) is curious.
29. Barnouw 1980 argues a continuity between them.
31. Cf. Descartes: 'causa sive ratio' (*Replies to the Second Set of Objections*, Ax. I), Spinoza: 'causa seu ratio' (*Ethics*, I, 11).
31. See Hobbes 1660: 'Qui figuras definiunt, Ideas, quae in animo sunt, non ipsa corpora respiciunt' (p. 87); 'Divisio est opus intellectus, intellectu facimus partes ... Idem ergo est partes facere, quod partes considerare' (p. 56).
32. So Descartes see the exposition in Cassirer 1922, I, pp. 469ff.
33. The question of Hobbes's intellectual affiliations is more complicated than this, of course, but a more complete treatment would not displace the point made. Such a treatment would have to take into account Hobbes's relation to Bacon whose well-known thesis 'scientia et potentia in idem coincident' is echoed almost to the word in Hobbes (1656, Ch. I, §6). It would also have to treat the much neglected topic of Locke's approximately contemporary account of morals as a demonstrative science, based on his distinction

between 'ectypes' and 'archetypes'. On the whole matter see his *Essay*, II.xxxi.13 & 14, III.xi.15–17, IV.iii.18, IV.iv.5–7, IV.xii.8. On more general related questions see Baruzzi 1973.
34. The following approach to the Galilean revolution has been decisively influenced by, among others, the following: Böhme, van den Daele, Krohn 1977, Cassirer 1922, Lewin 1935, Mittelstrass 1970 (see also the report on this in his 1972).
35. See the excellent Tetens 1987 affiliated to Dingler, esp. 1928 and 1952. There is a convergent treatment from the totally different perspective of Marxism in Raphael 1974, pp. 94f.
36. I take this formulation from Cassirer 1959, 406ff.
37. It is hardly possible to exaggerate the significance of this breaking of the nexus between science and what is actually the case at certain places and times: echoing Hilbert's famous remark about Cantor's work on the infinite, it created a paradise of free theorising from which nothing has been able to expel science. In particular it shows how conservative, how divorced from the meaning of the real character of the new science was Hume and his 'constant conjunction' theory of cause. For explicit statements by Galileo on the purely hypothetical character of theorising as such see, for example, the *Two New Sciences*, Day III, first paragraph of 'On Naturally Accelerated Motions' and the closely related letter to P. Carcavy of 5 June 1637 (Galileo's *Opere*, 17:90).
38. It is also very probably a result of something like the following line of reasoning. (a) A great deal of Vico's work on history is very valuable. (b) Vico's work on history is, partly anyway, founded on VP (that is, the latter is a necessary condition for the former, so that the former is a sufficient condition for the latter). (c) Therefore, from (a) and (b) it follows that VP is worthy of endorsement. But though (a) is true, (b) is not: VP is an extraneous metaphysical gloss on the historical work as such. So the argument gives no grounds for believing (c). – It might be added here that, contrary to not a little loose talk about the relation of VP to Marxism, there is not a single passage in Marx's writings that can be adduced as even the most indirect support for this. Of the three references to Vico in the Marx-texts now available, two of them (letters to Lassalle and to Engels of 28 April 1862) are quite brief remarks on particular historical points, and the third (in the course of the fourth footnote to Ch. XV the English edition of the first volume of *Capital*) simply records a banality.
39. It may be remarked that this sort of confusion between two or more intellectual formations is by no means a rare phenomenon in the history of ideas, and sometimes works for good, sometimes for ill. But it has not, so far as I am aware, been given a name, and I want to propose one here, taken from the distant domains of mineralogy and geology. These sciences are familiar with the phenomenon of minerals that have the crystal form of one species but the chemical composition of another. (Examples are malachite/cuprite, barite/quartz, limonite/pyrite.) The one is called a 'pseudomorph' of the other, and the two are obviously easily confused in the absence of deeper, chemical analysis. The analogous situation, where the same or very similar *words* are attached to different *concepts* (and similar *sentences* to different *propositions*), leading to confusion between the one and the other, I propose to call 'cognitive pseudomorphs'. (After writing the preceding it occurred to me that the latter was probably the effect of a subconscious memory of the general historical notion of 'pseudomorph' in Spengler 1926–28, where it is explicitly introduced at I, p. 189. Examples will be found at I, pp. 209–12, 214, 216, 228, and II, pp. 74, 189, 190, 191, 192, 200, 210, 211, 256–58, 349, 480n.)
40. The historical fate of the idea of linking knowing and 'making' is an extremely interesting, significant and complex one, which has not yet even been sketched comprehensively by anyone. (The best available account is a very incomplete and in some respects misleading one in Löwith 1968, pp. 19ff.) Some brief contributions to this outstanding problem follow. What would seem to be the first immediate response to Hobbes's 'genetic' account was Spinoza's treatment of definition in his *De intellectus emendatione*, §§95–

97 (and cf. §§71, 72) – Bruders paragraphing – on which, and other relevant matter, see McKeon 1930. Leibniz also gives a substantially similar treatment. (Relevant passages in his works are scattered. An excellent overview is given in Cassirer 1902, pp. 113ff.) It surfaces again in a central way in Kant's epistemology, first in his theory of mathematics, and second in his account of empirical knowledge. There can be little doubt, in the light of the well-known passage in the preface to the second edition of the *Critique of Pure Reason* (B xiii–xiv) especially, that he has a crucial affiliation with the Galilean experimental revolution. (See also *Critique of Judgement*, §75.) That it is in no way a Vichian idea – as was suggested by Jacobi 1816, pp. 352f whose remarks are an excellent example of the working of a 'pseudomorph (see the preceding note) – is shown decisively by one of his private notes: 'We do not really understand anything but what we can at the same time make, *if the materials for it were given to us*' (No. 395 of the *Reflexionen Kants zur kritischen Philosophie*, ed. B. Erdmann, Leipzig, 1882, my emphasis, WAS). But at a deeper level Kant of course introduces the central idea of a construction of objects as such (of the objects that are merely *re*constructed in experiment) in terms of the 'transcendental subject'. Even though the latter still 'works on' a given material, the way had been opened to a fateful development. Through Jacobi and Fichte as historical-systematic intermediaries Hegel arrived at the program of submitting the element of givenness (objecthood as such) in knowledge to a 'final solution' with the idea of what exists being considered as the self-constitution of the Absolute Subject, the 'being', so to speak, of the latter being conceived as its own eternal 'becoming'. Thus he arrives at an intellectually 'iridescent' species of secularised version of Vico's – and the Christian – conception of the oneness of God's creating and knowing the world, though this time *via* a rich series of particular insights. After Hegel the connecting of knowing and making turns up in all sorts of places (e.g. in Nietzsche), but three particularly significant places may be mentioned: in Marxism (see especially Engels 1888, Pt II, especially p. 347), Dewey (see especially his great, at the present time mostly forgotten work 1938, on the general idea of which in relation to both Vico and Hobbes see Child 1953, and in a broader context Kannegiesser 1977), and Dingler (see the references and also the related historical survey in Klüver 1977, as well as the Dingler-inspired interpretation of Hobbes in Weiss 1983 and other writings referred to here). See also the superb Gehlen 1988. Finally, it may be noted that I have not so far mentioned the related 'production' theory of knowledge, stemming, approximately, from the work of Louis Althusser, on which see Suchting 1986, especially pp. 16ff.

41. For Vico genuine science deals with the eternal, the unchangeable, the indubitable and is thus identical with metaphysics (e.g. 1988, pp. 66, 67, 69, 77, 92) and indeed a thoroughly idealist metaphysics (1988, pp. 93f and 1963, pp. 138f). Of the epoch-making breakthroughs of early modern scientific work, he is opposed to analytic geometry (1988, pp. 59f; 1963, pp. 144f), his theoretical physics is completely backward-looking (physical forces are seated in metaphysical ones according to 1988, p. 59, motion is to be understood in terms of *conatus*, anchored in God, pp. 78f, and he is therefore opposed to the very idea of a principle of inertia, pp. 80ff). And as to experiment the partisan of the *verum = factum* principle writes the following in the review of his life (he refers to himself in the third person): 'A short time after this he learned of the growing prestige of experimental physics ... but, profitable as he thought it for medicine and spagyric [alchemy – WAS], he desired to have nothing to do with this science. For it contributed nothing to the philosophy of man and had to be expounded in barbarous formulas, whereas his own principal concern was the study of Roman laws, the main foundations of which are the philosophy of human customs and the science of the Roman language and government, which can only be learned in the Latin writers.' Vico 1963, p. 128.

42. This is not, of course, to say that ordinary, everyday perception may not be pervasively understood in certain ways so customary that may be called, following Feyerabend 1975, p. 73, 'natural interpretations'. (An obvious example is the spontaneous geocentric view of the phenomena of 'sunrise' and 'sunset'.) But this does not entail that the perceived

situation itself is somehow subjectively 'constructed'. Such 'natural interpretations' can be made explicit and rejected, the perceived situation remaining the same. On the whole matter see also Husserl 1970a, esp. §9.

REFERENCES

Althusser, L.: 1969, *For Marx*, Allen Lane, The Penguin Press, London.
Althusser, L. and Balibar, E.: 1970, *Reading Capital* New Left Books, London.
Austin, J. L.: 1962, *Sense and Sensibilia*, Clarendon Press, Oxford.
Barnouw, J.: 1980, 'Vico and the Continuity of Science: The Relation of His Epistemology to Bacon and Hobbes', *Isis* **71**, 609–620.
Baruzzi, A.: 1973, *Mensch und Maschine. Das Denken sub specie machinae*, W. Fink, Munich.
Berlin, I.: 1976, *Vico and Herder. Two Essays in the History of Ideas*, Chatto & Windus, London.
Blumenberg, H.: 1987, *Das Lachen der Thakerin. Eine Urgeschichte der Theorie*, Suhrkamp, Frankfurt/M.
Böhme, G. (ed.): 1976, *Protophysik. Für und wider eine konstruktive Wissenschaftstheorie der Physik*, Suhrkamp, Frankfurt/M.
Böhme, G., van den Daele, W. and Krohn, W.: 1977, *Experimentelle Philosophie. Ursprünge autonomer Wissenschaftsentwicklung*, Suhrkamp, Frankfurt/M.
Butts, R. E. and Brown, J. (eds.): 1989, *Construction and Science. Essays in Recent German Philosophy*, Academic Publishers, Dordrecht/Boston/Lancaster.
Cassirer, E.: 1902, *Leibniz's System in seinen wissenschaftlichen Grundlagen*, N. G. Elwert, Marburg.
Cassirer, E.: 1922, *Das Erkenntnisproblem in der Philosophie und Wissenschaft der neueren Zeit* (third edition) 2 vols B. Cassirer, Berlin.
Cassirer, E.: 1923, *Substance and Function* in *Substance and Function and Einstein's Theory of Relativity* (1910/1921) Open Court Publishing Co., La Salle, Ill.
Cassirer, E.: 1959, *Philosophy of Symbolic Forms*, Volume 3: *Phenomenology of Knowledge* (1929) Yale University Press, Yale.
Cassirer, E.: 1985, 'Die Sprache und der Aufbau der Gegenstandswelt' (1932), in Orth, E. W. and Krois, J. M. (eds.), E. Cassirer, *Symbol, Technik, Sprache. Aufsätze aus den Jahren 1927–1933* F. Meiner, Hamburg.
Child, A.: 1953, 'Making and Knowing in Hobbes, Vico and Dewey', *University of California Publications in Philosophy* **16**, 271–310.
Collingwood, R. G.: 1939, *An Autobiography*, Clarendon Press, Oxford.
Dewey, J.: 1938, *Logic, the Theory of Inquiry*, Holt, Rinehart and Winston, New York.
Dingler, H.: 1928, *Das Experiment. Sein Wesen und seine Geschichte*, Ernst Reinhardt, Munich.
Dingler, H.: 1938, *Die Methode der Physik*, Ernst Reinhardt, Munich.
Dingler, H.: 1952, *Ueber die Geschichte und das Wesen des Experimentes* Eidos-Verlag, Munich.
Dingler, H.: 1969, *Die Ergreifung des Wirklichen Kapitel I–IV*, Suhrkamp, Frankfurt/M.
Dingler, H.: 1987, *Aufsätze zur Methodik*, Felix Meiner, Hamburg.
Engels, F.: 1970, *Ludwig Feuerbach and the End of Classical German Philosophy* (1888) in Karl Marx and Friedrich Engels, *Selected Works in Three Volumes* Volume 3, Progress Publishers, Moscow.
Feyerabend, P. K.: 1975, *Against Method. Outline of an Anarchistic Theory of Knowledge*, New Left Books, London.

Fritz, K. von: 1964, *Die ἐπαγωγή bei Aristoteles*, Verlag der Bayerischer Akademie der Wissenschaften, Munich.
Gaukroger, S.: 1986, 'Vico and the Maker's Knowledge Principle', *History of Philosophy Quarterly* 3, 29–44.
Gehlen, A.: 1988, *Man. His Nature and Place in the World* (1972), Columbia University Press, New York.
Glasersfeld, E. von: 1989, 'Cognition, Construction of Knowledge, and Teaching', *Synthese* 80, 121–140.
Hobbes, T.: 1839–45, *The English Works*, W. Molesworth (ed.), John Bohn, London.
Hobbes, T.: 1839–45a, *Opera philosophica, quae latine scripsit, omnia*, W. Molesworth (ed.), John Bohn, London.
Hobbes, T.: 1651, *Leviathan*, in Hobbes 1839–45, Vol. III.
Hobbes, T.: 1656, *Elements of Philosophy. The First Section, Concerning Body*, in Hobbes 1839–45, Vol. I.
Hobbes, T.: 1656a, *Six Lessons to the Savilian Professors of the Mathematics . . .*, in Hobbes 1839–45, Vol. VII.
Hobbes, T.: 1658, *Elementarum philosophiae sectio secunda, de Homine*, in Hobbes 1839–45a, Vol. II.
Hobbes, T.: 1660, *Examinatio et emendatio mathematicae hodiernae*, in Hobbes 1839–45a, Vol. IV.
Husserl, E.: 1970, *Logical Investigations*, Volume I (1913), Routledge and Kegan Paul, London.
Husserl, E.: 1970a, *The Crisis of the European Sciences and Transcendental Phenomenology. An Introduction to Phenomenological Philosophy* (1954) Northwestern University Press, Evanston.
Jacob, M. C.: 1988, *The Cultural Meaning of the Scientific Revolution*, Knopf, New York.
Jacobi, F.: 1816, 'Von den göttlichen Dingen und ihre Offenbarung' (1811), in *Werke*, Vol. III, G. Fleisher, Leipzig.
Kannegiesser, H. J.: 1977, *Knowledge and Society*, Macmillan, Melbourne.
Klüver, J.: 1977, *Operationalismus. Kritik und Geschichte einer Philosophie der exakten Wissenschaften*, Friedrich Frommann Verlag, Stuttgart-Bad Cannstatt.
Lefèvre, W.: 1978, *Naturtheorie und Produktionsweise. Probleme einer materialistischen Wissenschaftsgeschichtsschreibung. Eine Studie zur Genese der neuzeitlichen Naturwissenschaft*, Luchterhand, Darmstadt and Neuwied.
Lewin, K.: 1935, 'The Conflict between Aristotelian and Galilean Modes of Thought in Contemporary Psychology', in K. Lewin, *A Dynamic Theory of Personality*, McGraw-Hill, New York, pp. 1–42.
Löwith, K.: 1968, *Vicos Grundsatz:verum et factum convertuntur. Seine theologische Prämisse und deren säkulare Konsequenzen*, Carl Winter Universitätsverlag, Heidelberg.
Malcolm, N.: 1958, 'Knowledge of Other Minds', *Journal of Philosophy* 55, 969–978.
Mead, G. H.: 1934, *Mind, Self and Society from the Standpoint of a Social Behaviourist*, University of Chicago Press, Chicago.
Mittelstrass, J.: 1970, *Neuzeit und Aufklärung:Studien zur Entstehung der neuzeitlichen Wissenschaft und Philosophie*, W. de Gruyter, Berlin and New York.
Mittelstrass, J.: 1972, 'The Galilean Revolution. The Historical Fate of a Methodological Insight', *Studies in History and Philosophy of Science* 2, 297–328.
McKeon, R.: 1930, 'Causation and the Geometric Method in the Philosophy of Spinoza', *Philosophical Review* 39, 178–189, 275–296.
Pears, D.: 1988, *The False Prison. A Study in the Development of Wittgenstein's Philosophy*, Volume 2, Clarendon Press, Oxford.
Raphael, M.: 1974, *Theorie des geistigen Schaffens auf marxistischer Grundlage*, S. Fischer, Frankfurt/M.
Russell, B.: 1956, 'Mathematics and the Metaphysicians' (1901/1918), in J. R. Newman (ed.), *The World of Mathematics*, Simon and Schuster, New York.

Sellars, W.: 1956, 'Empiricism and the Philosophy of Mind', in H. Feigl and M. Scriven (eds.) *Minnesota Studies in the Philosophy of Science*, Universiy of Minnesota Press, Minneapolis, Vol. I, pp. 253-329.
Spengler, O.: 1926-28, *The Decline of the West* (1923) 2 vols., Knopf, New York.
Stove, D.: 1991, *The Cult of Plato and Other Philosophical Follies*, Basil Blackwell, Oxford.
Suchting, W.: 1986, *Marx and Philosophy. Three Studies*, Macmillan, London.
Tetens, H.: 1987, *Experimentelle Erfahrung. Eine wissenschaftstheoretische Studie über die Rolle des Experiments in der Begriffs-und Theorie-Bildung der Physik*, Felix Meiner, Hamburg.
Vico, G.: 1988, *On the Most Ancient Wisdom of the Italians Unearthed from the Origins of the Latin Language* (1710), Cornell University Press, Ithaca and London (1744).
Vico, G.: 1961, *The New Science* (1744), Anchor Books, Garden City, New York.
Vico, G.: 1963, *Autobiography*, Great Seal Books, Ithaca.
Vygotsky, L. S.: 1962, *Thought and Language*, MIT Press, Cambridge, Mass.
Weiss, U.: 1983, 'Wissenschaft als menschliches Handeln. Zu Thomas Hobbes' anthropologischer Fundierung von Wissenschaft', *Zeitschrift für philosophische Forschung* **37**, 37-55.
Wertsch, J.: 1985, *Vygotsky and the Social Formation of Mind*, Harvard University Press, Cambridge, Mass.
Williams, R.: 1983, *Keywords. A Vocabulary of Culture and Society*, Fontana Paperbacks (Flamingo), London.
Wittgenstein, L. 1958, *Philosophical Investigations* (2nd ed.), Basil Blackwell, Oxford.
Zagorin, P.: 1984, 'Vico's Theory of Knowledge: A Critique', *Philosophical Quarterly* **34**. 15-30.

Constructivism Reconstructed:
A Reply to Suchting

ERNST VON GLASERSFELD

37 Long Plain Road RFD3, Amherst, MA 01002, USA

Under ordinary circumstances, an author faced with a long and detailed critique that dismisses his paper as mostly unintelligible and for the rest simply confused, should be devastated, if not silenced for ever. But the circumstances in the case at hand are not ordinary. W. A. Suchting calls his effort *Constructivism Deconstructed* and within the first six pages he provides an instructive example of his very own method of deconstruction. Though the traditional philosopher's style of writing might deceive the unwary reader, Suchting's method has the virtue of simplicity and will undoubtedly be effective – especially with readers who do not have access to the original piece.

The term *deconstruction* has mostly been used in the context of texts or works of art, and it is clear that this deconstructing procedure cannot be quite the same as in the dismantling of a car engine or a chain saw. Yet, in the case of a text, just as with the mechanical items, it must begin from the surface, loosening, as it were, a screw here and there to separate the parts in order then to examine both them and the relations that are supposed to justify the structure as a whole. The surface of texts are printed words, and it is therefore quite proper that Suchting, when he begins to deconstruct, focuses his attention on a string of words on the first page of my paper (v. G., 1989).[1] But Texts, too, have parts, e.g., sentences, whose integrity must be respected, at least at the beginning of any deconstruction. And there is an immediate surprise. The fifteen words Suchting quotes are the first of my abstract, not of my paper (Suchting, p. 3). But this is a minor consideration. More importantly, these fifteen words are the beginning of a sentence that runs as follows:

> The existence of objective knowledge and the possibility of communicating it by means of language have traditionally been taken for granted by educators. (v. G., 1989, p. 121)

Suchting, however, presents my fifteen words preceded by the phrase: 'The paper opens by saying that what constructivism denies is . . .' and demonstrates with much logical acumen that this piece of the text can be 'excised' because the negation makes part of it meaningless. He then proceeds to use this result as launching pad for an erudite critique of my use of the word 'objective' (both in the quoted excerpt and on the last page of my essay) and to show that I am not only confused but indeed ignorant of the word's meaning.

Struck by the obvious power of an opening move that consists in constructing yourself what you then deconstruct, it is not surprising that I went on reading with the greatest interest. And sure enough, though undoubtedly the most transparent instance, the gambit is used many more times in Suchting's critique. But as I can neither hope that many readers will consult the original text nor want to quote myself in every paragraph, I shall give another example.

Suchting picks the phrase 'the world as it is' from a footnote and declares: 'The problem with expressions such as these in the context of what is being denied is that, taken by themselves, they are vacuous' (p. 5). The oddity here is that my phrase does not stand in the context of negation but in a note that tries to lay out the ambiguity in the current usage of the word 'objective'; and the note does not stand by itself but is attached as amplification to a sentence of the text (v. G., 1989, p. 124) that contains the expression: 'objective representation of an observer-independent world'.

On page 7, my statement that constructivists 'posit knowledge as a mapping of what, in the light of human experience, turns out to be feasible' is turned into 'it is said (135) that knowledge qua feasible is a "mapping"', and a couple of lines further Suchting transforms this into: '... and "mapping" is what replaces "description"/"representation"/"correspondence"'. Three pages later (9–10), this spurious interpolation is 'deconstructed' in a lengthy passage to bring to light that, deep down, constructivism after all *does* dabble with a 'correspondence theory of truth'.

Professing disappointment at not being able 'to "construct" a clearer picture of constructivism than is available in the paper under examination' (p. 17), Suchting quotes what Russell once (I believe rather flippantly) said of mathematics, namely that it is a subject 'in which we never know what we are talking about, nor whether what we are saying is true' (p. 17). Apparently my critic is still asking whether constructivism is 'true'. This is precisely the reaction of Vico's contemporary (see v. G., 1989, p. 123), who wanted Vico to *prove* the theory of knowledge he had been expounding and who thus showed that he was disregarding a core posit of the theory, namely that knowledge can never be considered true in the conventional sense (i.e., correspond to an observer-independent reality) because it is *made* by a knower who does not have access to such a reality.

The next paragraph (p. 17) confirms that impression. 'Knowledge', he says, 'in any ordinary, understandable sense of the word requires something other than the knowledge of which the latter can be said to be knowledge'. I understand this to mean that knowledge must be a representation of something else, and this, indeed, is the conventional meaning of the word. It seems a trifle wayward, however, to insist on this meaning in the 'examination' of a paper whose author announces as early as in the abstract that he intends to use 'knowledge' in Piaget's *adaptational* sense to refer to those sensory-motor actions and conceptual operations that have proved viable in the knower's experience.

Concerning Vico's Principle (VP), Suchting says '... according to VP we cannot be said to have access to "the true" *and hence to knowledge*' (p. 18, my emphasis). Earlier, however, he reported Vico's thesis number 1 (p. 15) which states that a knower can know only what the knower has made, i.e., the *factum*, and therefore (thesis 2, p. 16) a human knower cannot know God's natural world. I would suggest that this is a flat contradiction – unless we admit that 'to know' can be used in two different senses: Vico's sense and that of traditional, representationist epistemology. Suchting unfortunately persists in disregarding the explicit announcement that constructivism accepts Vico's theses and also agrees with Vico's revolutionary declaration that 'the true' (verum) is what has been 'made' (*factum*).

It would take far more space than is available here to deconstruct Suchting's 'more "relaxed" version of VP' (p. 19) and, at least in my view, it would not add much to the discussion of constructivism. The reason is that Suchting characterizes this version by saying that, in it, 'the actual 'creation' of the elements and relations are not in question'. I take this to mean that they are considered ready-made or 'given' raw material for any construction. This would be incompatible with a modern constructivism that sprang, as I have often said, from the immensely rich work of Piaget concerning precisely the child's *construction* of concepts and conceptual relations. Personally, I would not want to call this 'creation' because, although there is a certain novelty about the products, they are not conjured up *ex nihilo*.

Discussing a passage of my paper that mentions the child's construction of the notion of 'others', Suchting remarks: '... the consequence is that (a) the subject always experiences/knows other subjects and also (b) that there is a time ... when the subject does *not* experience/know others'. He sees this as an example of 'self-contradiction' and he is right – but the synonymy indicated by the slash between 'experience' and forms of the verb 'to know' is Suchting's construction, not mine.

If knowledge, as I proposed (v. G., 1989, p. 124), is intended to refer 'to conceptual structures that epistemic agents, given the range of present experience within their tradition of thought and language, consider *viable*', it should be clear that 'to know' (in this restricted sense) cannot be equated with 'to experience'. Indeed, as I say later, 'the most frequent source of perturbations for the developing cognitive subject is the interaction with others' (p. 136), and given that 'perturbation' was defined as the failure of a 'scheme' (a piece of knowledge that, so far, has proven viable) it follows that perturbations must be part of experience but are not in themselves knowledge (though they may lead to new knowledge via an accommodation).

Concerning my model of how the child might construct 'objects' and 'others', Suchting refers to Wittgenstein, who led him to the conclusion 'that there is no sensation-language (as it may be called) independent of and prior to object-language, ...' (p. 22). This would seem to be an existential statement and, if it is, it is out of place in this context. I use the term 'model' in the sense that was introduced by cybernetics, namely for a tentative construct, physical or conceptual, as stand-in for an inaccessible process that could yield a known result. In the case in point, namely the construction of an object-concept,[2] the Piagetian model does not involve anything like a 'sensation-language' but only the organism's capability of making distinctions (say, in its visual field), acting upon the made distinctions, and, as I laid out in my paper, the ability to establish recurrences (of distinctions made). This model works remarkably well in our dealings with animals – for instance bees, who readily associate a particular color with the entrance to *their* hive and do not try to enter a hive marked with another color. It is a perfectly *viable* model and the fact that we may have not the foggiest idea of what might actually be going on in a bee's head in no way diminishes that viability.

As Suchting mentioned earlier in his deconstruction of 'objectivity', the Galilean transformation-equations (constituting a model of kinematics) have been partially replaced by Einsteinian equations. The Piagetian model for the construction of object-concepts may sooner or later be replaced by a more comprehensive one, but at the moment I know of no better.

Concerning the notion of 'self', my critic raises the point, 'strongly argued by many philosophers', that it is 'the *community* of subjects/others that constitutes individual subjects, or, better, that self and other are correlative' (p. 22). This may well be plausible on a later level of development where all sorts of abstract properties are conceptualized, but I was focusing on the beginnings. If Suchting has had the opportunity to spend a certain amount of time with an infant during its first year, he might have observed that there is a moment when the infant notices that biting its own toe is different from biting the rattle or the comforter. This, I suggest, *pace* Hegel and the other philosophers Suchting had in mind, is one of the first steps in the subject's separation of the 'self' from the rest of the experiential world.

Though he cites Kant when it suits him, he chooses to disregard the Kant quotation (v. G., 1989, p. 130) from which I explicitly derive my development of the notion of 'other'. Hence I am not too worried about the 'gap' in my exposition (quoted from that very page) and which, in his romantic vein, the critic characterizes as a 'dark unbottom'd infinite abyss' (p. 23).

Many of the arguments Suchting proffers against my brief account of early language acquisition (pp. 23–28) might be cogent had I not first taken some trouble to explain the Piagetian model of learning, the terms of 'assimilation' and 'accommodation', and the constructivist notion of viability. Taken together, I had hoped, these explanations would make it apparent that the use of words is learned as an instrumental activity and that the 'interpersonal fit' that so profoundly disturbed my critic would be understood as gradually deriving from the accommodations following upon perturbations caused by the learner's failure to achieve his or her goal.

Prefacing the last section, a most interesting comparison of Vico and Hobbes (pp. 28–34), Suchting describes the wonderfully benign, olympian position from which philosophers, he thinks, should approach the philosophical utterances of others. There he speaks of the need of 'sympathetic reading', of asking 'whether we had really understood the point of his (the other's) words or behaviour, and whether, this having been grasped, his words or behaviour might not seem quite intelligible to us' (p. 28). It warmed my heart to see that he is able to adopt such an admirable attitude, even though he apparently reserves it for the dead and treats the living very differently.

The last item I want to bring up is Suchting's uncovering of the fact that constructivism is a form of empiricism (pp. 11 and 34). Given that the paper he is criticizing repeatedly draws attention to the *instrumentalist* character of the ideas presented (v. G., 1989, pp. 28, 135, 136), his discovery is not much of a revelation. His attempt, however, to translate the terms I used into what he considers the dictionary of 'a standard, middle-of-the road, more or less recent empirical position' (p. 11) shows not only that I have totally failed to get my meaning across, but also that Suchting is not above playing fast and loose with meanings in general. His equation '"adapted" = "confirmed"' deprives the biological term of its instrumentalist core meaning, namely *to have the wherewithal to survive* or, as I might say, to be *viable*. The equation '"scheme" = "theory"' disregards the fact that I explicitly used the term 'scheme' in the Piagetian sense of 'scheme of action' or 'scheme of operation' (adopting a computer metaphor, schemes could be seen as programs or subroutines). And I doubt that many standard empiricists would agree to use the term 'theory' for goal-directed action patterns, which is what a large part of children's and adult's schemes are. No more acceptable is the jump from one domain to another when 'perturbation' is translated as 'disconfirmation'/'falsification', because the latter involves, even for empiricists, some notion of truth, whereas the former is tied to the notion of viability. Instead of bothering Horace, I would quote the old Italian saying: *traduttore traditore*.

Nevertheless, I, too, consider constructivism an offspring of subjective empiricism. It makes no secret of having adopted Locke's much neglected insight that the source of complex ideas is the mind's reflection upon its own operations,[3] Berkeley's principle that it is 'impossible for us to conceive a likeness except only between our ideas',[4] and it not only accepts Hume's thesis that the causal relation requires an experiencer,[5] but extends this requirement to all relational notions. But since constructivism is explicitly instrumentalist, it holds that all this conceptual construction is carried out not for the sake of representational knowledge of a 'given' world, but to enlarge the map of viable pathways in the world constituted by the subject's experience.

In his conclusion, Suchting asserts that constructivism claims to be 'the New Age in philosophy of science, . . .' (p. 35). Once more he is off the mark, because he could have read on the second page of my paper that I consider my view 'not as a new invention but rather as the result of pursuing suggestions made by much earlier dissidents' (v. G., 1989, p. 122). Indeed, it might have been this very sentence that prompted him to call constructivism a gallimaufry – an unkind epithet, but not inappropriate, because radical constructivism certainly (and explicitly) combines several ideas that the Western philosophical establishment left by the wayside in the course of history.

Underlying most of Suchting's argumentation seems to be the tacit assumption that Truth, if not self-evident, can at least be recognized when it is professionally presented. For a constructivist this is an illusion. From its perspective, Truths are replaced by viable models – and viability is always relative to a chosen goal. To end my reply I want to quote something I only recently read: 'Theory can simply continue doing what all discursive practices do: attempt to persuade its readers to adopt its point of view, its way of seeing texts and the world.'[6] – The failure to persuade a critic at least to see one's point of view is always painful; but it also generates a salutary perturbation that may lead one to find better ways of expression.

NOTES

1. Glasersfeld, E. von: 1989, 'Cognition, Construction of Knowledge, and Teaching', *Synthese* **80**, 121–140.
2. Note that in Piaget's theory the development of an 'object-concept' is a necessary but preliminary step to the construction of 'object permanence'.
3. Locke, J.: 1690, *An Essay Concerning Human Understanding*, Book II, Chapter 1, Section 4.
4. Berkeley, G.: 1710, *A Treatise Concerning the Principles of Human Knowledge*, Part 1, Section 6.
5. Hume, D.: 1750, *An Enquiry Concerning Human Understanding*, Essay VII, Part I.
6 Mailloux, S.: 1985, 'Truth or Consequences: On Being against Theory', in W. J. T. Mitchell (ed.), *Against Theory*, University of Chicago Press, Chicago, pp. 65–71.

Constructivisms and Relativisms: A Shopper's Guide

MARK H. BICKHARD

Department of Psychology, 17 Memorial Drive East, Lehigh University, Bethlehem, PA 18015, USA

ABSTRACT. Diverse forms of constructivism can be found in the literature today. They exhibit a commonality regarding certain classical positions that they oppose – a unity in their negative identities – but a sometimes wild multiplicity and incompatibility regarding the positive proposals that they put forward. In particular, some constructivisms propose an epistemological idealism, with a concomitant relativism, while others are explicitly opposed to such positions, and move in multifarious different directions. This is a potentially confusing situation, and has resulted in some critics branding *all* constructivisms with the charge of relativism, and throwing out the baby with the bath water. In addition, since the epistemological foundations of even non-relativist constructivisms are not as familiar as the classical positions, there is a risk of mis-interpretation of constructivisms and their consequences, even by some who endorse them, not to mention those who criticize. Because I urge that some version of constructivism is an epistemological necessity, this situation strikes me as seriously unfortunate for philosophy, and potentially dangerous for the practice of education.

Key words: realism, idealism, empiricism, innatism, constructivism, relativism, representation, epistemology, scaffolding

Diverse forms of constructivism can be found in the literature today. They exhibit a commonality regarding certain classical positions that they oppose – a unity in their negative identities – but a sometimes wild multiplicity and incompatibility regarding the positive proposals that they put forward. In particular, some constructivisms propose an epistemological idealism, with a concomitant relativism, while others are explicitly opposed to such positions, and move in multifarious different directions. This is a potentially confusing situation, and has resulted in some critics branding *all* constructivisms with the charge of relativism, and throwing out the baby with the bath water (Bickhard, 1993b). In addition, since the epistemological foundations of even non-relativist constructivisms are not as familiar as the classical positions, there is a risk of mis-interpretation of constructivisms and their consequences, even by some who endorse them, not to mention those who criticize.

Because I urge that a constructivism is an epistemological necessity, this situation strikes me as seriously unfortunate for philosophy, and potentially dangerous for the practice of education (Bickhard 1995). The issues that frame epistemological constructivisms and their alternatives are of fundamental importance, but they are not simple and cannot be safely accepted, dismissed, or criticized in an all-or-nothing fashion.

My primary aim, then, is to outline what I take to be some of the most important issues framing this debate, with a secondary focus on a few particular versions of constructivism. I strongly endorse a rejection of idealism and relativism, but will argue that rejecting constructivisms *in toto* on these grounds has the ironic result of committing to the problems that yield idealisms and relativisms in the first place. Rejecting constructivism, then, is precisely the wrong 'solution'.

A FRAMEWORK OF ISSUES

Realism and idealism

To begin an exploration of some of the framing issues, consider the classical opposition between realism and idealism. Realism in its classical forms begins with an assumption of a basic split between the knowing agent and the world. Both sides of this metaphysical division are taken for granted, and epistemological questions generally focus on how the epistemic agent – the knower – can know about the world. All classical attempts to answer this question about representation assume that representation consists of mental elements or events that are in some sort of representational correspondence with whatever they represent in the world. Those elements or events are taken to encode corresponding objects and properties in the world (Bickhard 1993; Bickhard & Terveen 1995).

Skepticism

A very large maze of venerable issues arise at this point (Bickhard 1993; Bickhard & Terveen 1995). One troublesome question that can be asked is how the knower can ever know whether or not his or her representations are correct. If I wish to check my representation that there is a desk in front of me, my only recourse would seem to be to again invoke my representation that there is a desk in front of me. Any such check seems to be unavoidably circular, and, therefore, no check at all. This is one of the core arguments of skepticism (Rescher 1980). It has not received adequate answers over the course of several millennia.

But, if there is no way to determine whether or not representations of the world are in error, then in what sense does it make sense to posit such a world at all? Perhaps all we *really* have is our representations, and there is no 'other end' of the presumed representational correspondences. Such a response to the apparent unanswerability of the skeptical arguments is a standard move into an idealism: the stance that our representations, be they individual or social, are all the reality there is.

Such an idealism, in turn, yields the apparently easy conclusion that 'anything goes'. If the only reality there is is constituted in our representations, and there are no rational criteria by which we can determine that some are correct and others incorrect, or even that some are better than others, then we arrive at a relativism. Relativisms are poisonous with respect to issues of morality and ethics, and can be psychotic with respect to issues of science and mathematics. If *anything goes* with respect to whether or not there is a desk in front of me, or whether or not there is a truck bearing down on me, then I have fallen into a potentially fatal psycho- or philosophico-pathology.

The argument that drives this move is the skeptical challenge to being able to check our representations. That challenge, in turn, depends on the standard conceptions of representations as being constituted as encoding correspondences. It is a challenge, one of many, to the presumed representational correspondences being able to work in the manner that they are supposed to work. It is frequently assumed that the source of the problems is the original division between knower and known, and that such a dualism must be transcended (e.g., Gergen 1995). I do not disagree with that position with respect to the standard way in which the dualism is understood, but I will argue that the shared assumption about representation is the more fundamental source of the dilemma and its disastrous consequences. Let us now turn to a classic division of assumptions about *how* those representational correspondences are supposed to work.

Empiricism and rationalism

Another of the many questions that can be raised about presumed encoding representations is how the epistemic agent is supposed to know what is even *supposed* to be on the other end of the correspondence – what is it that is supposed to be represented – prior to the skeptical issue of how it can be determined whether or not that representation is correct. Many representations are presumed to be constructed out of more basic representations – e.g., a desk representation might be some composite of representations of color, rectangle, legs, and so on. For such composite representations, issues of how we could know what is supposed to be represented devolve on the component representations – how do we know what *they* represent. Such decomposition of complex representations cannot proceed indefinitely, however, so there must be some level of non-decomposable atomic representations. The basic question of how we know what is supposed to be represented, then, has its ultimate focus on such presumed atomic representations.

There are two classes of classic answers to this question: 1) Knowledge of what is supposed to be represented by the foundational atomic representations enters into the mind along with the causal process – e.g., visual transduction in the retina – that generates the correspondences in the first place, or 2) Knowledge of what is supposed to be represented by the foundational atomic representations is already present in the mind. That is, the answers tend to fall into the classic empiricist or rationalist camps, or some mixture of them. There are, in turn, two versions of rationalism:

1) Basic representational knowledge is inherent in the nature of mind, or 2) It is resident in the genes. Contemporary versions of rationalism tend to advocate the innateness version – it's all in the genes (Bickhard 1991; Fodor 1981).

Note, however, that empiricism and rationalism share the assumption that representation is constituted as correspondence, and share the assumption that knowledge of what foundational representations represent – what's on the other end of foundational correspondences – *must come from somewhere*. Such knowledge (such *content* of the representations) must come from outside (empiricism) or inside (rationalism).

Emergence

This assumption, however, cannot be correct. If representation can exist only if it comes *from* somewhere, then representation could never have come into existence. Since representation presumably did not exist at the Big Bang, representation has to have come into existence at some point or points since then. Neither empiricism nor rationalism, and, therefore, classical realism, can account for such an emergence of representation. They render the existence of representation impossible.

A model of some phenomenon X must be consistent with X coming into being, with the emergence of X. If it is not, then that model *cannot* be a correct model of X. The skeptical arguments already demonstrate that there must be something wrong with standard models of representation. This impossibility for standard models of representation to account for the emergence of representation is another such demonstration. It points to a need for models of representation that could account for representational emergence. A model of representation that could account for emergence might also succeed in avoiding the pernicious dichotomy between classical realism, with all of its skeptical impossibilities, and idealism, with its nihilistic relativism. I will develop a model that I claim does precisely that.

Representation, action, and constructivism

The mind and mental phenomena have for millennia been studied within the assumption that consciousness is the proper locus of analysis, and with passive conceptions of vision as the metaphors for consciousness. This view yields a conception of knowledge that is sometimes characterized as 'knowing that' something is or is not the case – a propositional model of knowledge and representation. A century ago Charles Saunders Peirce introduced a shift to action as the proper locus for the study of mind and mental phenomena (Joas 1993; Rosenthal 1983). This is characterizable as a focus on 'knowing how', how to do something. The move from knowing that to knowing how, from correspondence-propositional models to pragmatic interactional models, I suggest, is the critical move for transcending the maze of dilemmas and paradoxes that were adumbrated above.

I do not wish to suggest that Peirce got it all right, nor those influenced by him (such as, for example, Heidegger (Joas, 1993) and Piaget). There are many details in a successful pragmatist model of representation, and many problems specific to such an approach that need to be addressed. But a pragmatist approach, assuming it is successful in these other respects, does offer a relatively easy avoidance of the classical impossibilities of the correspondence-encoding approaches. So, I will leave the details to elsewhere (Bickhard 1993; Bickhard & Terveen 1995). For current purposes, it suffices to indicate how problems of representational emergence might be solved or dissolved.

Classical models assume that representation is constituted as some sort of correspondence. That has not worked. Pragmatist approaches assume that representation is somehow emergent in systems of action and interaction. How could that be?

Evolutionary continuity. Note first of all that a locus of analysis in action and interaction honors a continuity in the animal kingdom, from the simplest to the most complex. Darwin highlighted this continuity – and opened up the consideration for Peirce. Without such a continuity, we are faced with an inexplicable saltation from animals that do not have propositional 'knowing that' consciousness to those that do (Churchland 1989).

Constructivism. If we construe representation as constituted by correspondences, we are tempted to assume that those correspondences are impressed into a passive mind, like forms into wax, or light transduced in the retina. If we assume that representation is an emergent of interactive systems, there is no such temptation to assume that the environment could impress the proper organizations of system into a system. Interactively successful system organization cannot be impressed into a system from its environment. System organization must be constructed by the system itself. *A pragmatic approach to knowledge and representation logically forces a constructivism.*

There is no temptation, for example, to assume that knowledge of skills must pre-exist somewhere and come into the organism acquiring those skills, nor that knowledge of how to survive as a frog must come into the frog genome from the environment or from pre-frog evolutionary sources. In both cases, the relevant knowledge is emergently constructed, either by the organism or over the course of evolution or both.

This relationship between interaction models and constructivisms, of course, is the proximal reason for my advocacy of constructivism, but the strength of that advocacy depends on the success of the underlying pragmatic framework in accounting for knowledge and representation. So, again, how could that be?

Interactive representation. Interactions sometimes succeed and sometimes do not. Whether or not they succeed depends in part on appropriate conditions holding in the environment with which the interactions are taking place. Engaging in a particular interaction, or indicating that a particular interaction would succeed if it were undertaken, presupposes that those appropriate conditions hold. They might not. If they do not, the interaction will fail, and that failure of the interaction falsifies that indication of presupposed conditions.

An indication of the appropriateness of an interaction, thus, has a truth value, and a truth value potentially (though fallibly) detectable by the system itself (Bickhard 1993; Bickhard & Terveen 1995). It captures the minimal characteristic of representation, of knowing that – it can be true or false. This, I argue, is the core sense in which a pragmatic approach can model representation.

Note that issues of emergence are trivial in this view: *any* appropriate organization of the system process will constitute representation. Therefore, the new construction of any such system organization will constitute *emergent* representation. Representation can be emergently constructed, constructed out of material that is not itself already representational.

This is the barest gesture toward a pragmatist or interactivist model of representation. There are at least three major classes of questions that need to be addressed: 1) How can such notions as 'indication', 'interaction type', and 'detection of success or failure' be modeled, without circularly invoking some other representational notions? 2) How can this minimal model of representation account for more familiar sorts of representations, such as of objects, and of abstractions, such as numbers? and 3) How can this model account for language? I will not address any of these sets of issues here (see Bickhard 1992b, 1993, 1995b; Bickhard & Terveen 1995). It suffices that there is some plausibility that a pragmatic approach is necessary to avoid classical aporias, and that a pragmatic approach forces a constructivism. We can now turn to looking at some of the varieties of constructivism, not all of which succeed in solving or dissolving these issues, and, therefore, not all of which are supported by these considerations.

SOME VARIETIES OF CONSTRUCTIVISM

Linguistic idealism

Kant argued against the empiricism of Hume that the mind provided its own organizing and framing principles for our representations. This constitutes a contributory function of the mind, but still leaves the mind essentially passive with respect to the world. Unlike pragmatist models, the notion of an intelligent plant – with no actions at all – is not an internal contradiction in Kantian terms.

Kant's own proposals concerning what the mind contributes to our cognitions have been largely abandoned. The discovery of non-Euclidean geometry, for example, made it clear that the mind does not impose Euclidean intuitions on our representations of space. Nevertheless, the idea that our understanding of the world is framed by prior conceptions has remained. One important descendent version holds that it is language that provides the framing conceptions (Gier 1981).

Construing language as the source of our representations of the world has at times been taken to the point of a linguistic idealism, in which our world is constituted by our language. Consider, for example, the hermeneutics of the later Heidegger and Gadamer: 'That which can be understood is language.' (Gadamer 1975, p. 432) 'Man's relation to the world is absolutely and fundamentally linguistic in nature.' (Gadamer 1975, p. 432) '... we start from the linguistic nature of understanding ...' (Gadamer 1975, p. 433) 'All thinking is confined to language, as a limit as well as a possibility.' (Gadamer 1976, p. 127). The details, of course, are complex and deserve much more attention, and, in fact, Gadamer wishes to deny the idealist stance that he seems to have committed himself too (Gadamer 1975, p. 496). Nevertheless, this historical development is 'just' a modern version of the move from naive realism to idealism, and has produced powerful contemporary versions of idealisms and positions committed to idealisms in spite of their own author's intentions (Gergen 1995; Rorty 1979). The nihilism and relativism that follow have even been acknowledged (Rorty 1987).

Note that such a linguistic idealism *is* a constructivism. Society constructs our representations, and, thus, our world, in the course of the evolution of language. What makes this position an idealism is that the notion of constraints on those constructions – of a world outside or independent of the constructions that nevertheless constrains the constructions – has been lost. Indeed, arguments for these positions generally proceed by recounting the historical failures to make good on correspondence notions of representation, failures to counter the skeptical arguments, and then shift to a social idealistic constructivism as if that is the only alternative (Gergen 1995; Rorty 1979, 1987). Such oscillations between realism and idealism are ancient, but they do not succeed in escaping the trap of the correspondence notions of representation that force the dichotomy in the first place. The nihilisms and relativisms that follow from such positions, however, are due to the constructions being unconstrained, not to the proposal that representations are constructed. That is, the nihilisms and relativisms are due to the idealism, not the constructivism.

Empiricist or rationalist 'constructivism'

There is, of course, a kind of constructivism that can be proposed, and has been proposed, even within an empiricism – or rationalism. This is the construction of composite representations out of the presumed atomic representations. For a time, there was even a kind of growth industry in Cognitive Science and Developmental Psychology in exploring such molecular constructions (Fodor, Bever, & Garrett 1974).

Such an endeavor, however, does nothing to solve the basic problem of the representational atoms themselves, and is inherently powerless to do so. Constructing representations out of atoms that are already themselves representations cannot address the origins of the atoms per se. Careful pursuit of such approaches has yielded bizarre *reductio ad absurdums* such as the implication that virtually all concepts that human beings are capable of having are already in the genes. After all, they could not have been created by any method known to contemporary cognitive science (Bickhard 1991; Fodor 1981).

It is clear that these versions of constructivism do not fall into the idealism error, but they avoid that error simply by remaining in the camp of some mixture of empiricism and innatist rationalism. They do not escape the fundamental problems.

Jean Piaget

Influenced by both Kant and Peirce, Piaget proposed a powerful version of constructivism based on pragmatic action, and coordinations among potential actions. Piaget was well aware of the classic battles between empiricism and rationalism, and felt that there had to be a 'third way', a way that avoided the problems of both. In particular, he focused on logical and mathematical necessity as a form of knowledge that could not be explained empirically: no matter how many times two pebbles plus two pebbles is observed to yield four pebbles, that does nothing to show that two plus two *necessarily* equals four. The number of planets is nine, but not necessarily so, while three times three is also nine, but, in this case, it is necessarily so.

Piaget's proposal for the origin of necessity was ingenious. As the mind constructs greater and greater ability to coordinate actions in some space of possible coordinations, it will eventually exhaust any bounded such space. At that point, the coordinations of which the mind is capable, at least in many cases, will form a mathematical group, or some other mathematical structure. In the emergence of such groups is the emergence of the necessity of all of those mathematical properties that groups (and other structures) necessarily have. Here is the *emergence* of mathematical necessity – it does not come in from the environment nor up from the genes.

Piaget's model is of a kind of inherent tendency in development. Any sufficiently powerful construction process will tend to exhaust bounded spaces of possible constructions, and fundamental new properties can come into being when such a space is subsumed in those constructions. Piaget looked to embryology for an analogy to such development – the tendency is inherent in the nature of construction – but it was quite explicitly no more than an analogy. The tendency for the emergence of necessary properties did not rest on anything in the genes beyond a sufficiently powerful construction process per se, nor on anything from the environment beyond sufficient stimulation to continue to 'feed' the processes of construction. Construing Piaget as a maturationist, or empiricist, or rationalist, then, is a serious misunderstanding (Kuhn 1992; Lourenço & Machado 1995).

I argue that Piaget did not ultimately succeed in his aims (Bickhard 1988; Campbell & Bickhard 1986). Nevertheless, he demonstrated the plausibility of constructing our familiar world of objects in space and time, related by causality, and amenable to mathematical analysis, within a framework of organizations of action and interaction. He demonstrated the plausibility of a pragmatist constructivism not only for frogs, for which a 'restriction' to 'knowing how' might be superficially plausible, but for the 'knowing that' world of human beings as well.

Popper and Campbell

Whether at the cultural and institutional level of science, or at the level of individual organisms, there can be no *a priori* guarantee that epistemic constructions about the world will be correct. Any such assurance constitutes prior knowledge, which begs the question of the ultimate origins of knowledge and representation. Epistemic constructions, then, must be trials that may or may not succeed – variations that may or may not be selected out. Pragmatic constructivism about the world must be a quasi-evolutionary process. Pragmatic constructivism forces an evolutionary epistemology (Campbell 1974).

Clearly we are not, in most circumstances, bereft of prior knowledge. Most of our problem solving trials are heuristically guided. The existence of such prior heuristic knowledge is of critical importance, and, in fact, imposes its own powerful constraints on the architecture of any epistemic system capable of such heuristics (Bickhard & Campbell, in press), but heuristic knowledge is itself knowledge, and its origins too must be accounted for. No model of the emergence of knowledge and representation can logically *require* prior knowledge or representation on pain of infinite regress, and the impossibility of emergence.

Of crucial importance to the current discussion is the point that no evolutionary epistemology can consistently be an idealism. In a consistent idealism, there can be no source of selections, no surprises. If there is nothing independent of the constructions, then there is nothing to select against, to resist, to surprise, those constructions. An evolutionary epistemology offers the possibility of accounting for the constructive emergence of representation without the 'free' unconstrained constructions of idealisms, and the consequent nihilisms and relativisms. This offer, of course, involves massive promissory notes concerning the specifics of representational and other normative emergences to be able to account for normative errors ranging from functional error to representational error to error of rationality to moral and ethical error. But a variation and selection constructivism does avoid the aporetic problems of not being able to account for any such emergence at all, in principle (Bickhard 1993, in preparation; Bickhard & Terveen 1995).

Von Glasersfeld

Von Glasersfeld's radical constructivism is a version of a pragmatic constructivism. It emphasizes the impossibility of obtaining correspondence knowledge about the world, and the ubiquity of considerations of pragmatic success. Under the rubric of 'viability', von Glasersfeld proposes a construction model subject to pragmatic selections. He emphasizes the ubiquity and necessity of feedback, the necessity of constructions in the light of such feedback, and the underdetermination – relative to fixed one-to-one correspondence models – that viability selections provide for constructions that survive those selections (von Glasersfeld 1979, 1981, 1995).

Such underdetermination could be read as 'no constraints', yielding an interpretation of radical constructivism as an idealism and consequent relativism. Clearly, this badly misses the point. Or, in an even worse interpretation, the focus on consequences could be construed as a verificationism, and radical constructivism could be read as a positivist empiricism. (A frequent mis-interpretive fate for pragmatism in this century.) Clearly, this egregiously misses the point. Radical constructivism may have its faults, but idealism and empiricism are not among them.

On the other hand, radical constructivism has not addressed the issue of the emergence of representation – more carefully, of precisely what viability feedback, especially error feedback, is and how it occurs. Von Glasersfeld has discussed notions of both representation and re-presentation, but the problem of how error, or 'non-viability', of representation is supposed to work, is not developed (von Glasersfeld, 1995). Radical constructivism thereby leaves this critical problem of representational error open, and, by default, potentially open to the same correspondence problems that radical constructivism seeks to avoid. Constructivism cannot ultimately simply avoid problems of representation, but if correspondence encodings are the temptation when representational issues are considered, then the classic framework of problems resurfaces. This lacuna in radical constructivism, I suggest, leaves it vulnerable to idealist and empiricist misinterpretations.

Constructivisms and relativisms

The varieties of constructivism are manifold. Some are idealist, some are even empiricist, but constructivism is not committed to either position. Constructivism is not committed, in fact, to any of the framework of classical problems generated by correspondence notions of representation. It offers a potentiality of avoiding the entire framework of realism-idealism, empiricism-rationalism, and all of its consequent aporias. In particular, it offers the possibility of accounting for representational emergence and for the possibility of organism detectable representational error.

Note that an evolutionary epistemological constructivism does postulate a 'realism'. It is not, however, a realism construed as the other end of representational correspondences – that is the view that generates the classical labyrinth of dead ends. Instead, it is a 'realism' construed as selections on constructions that are independent of those constructions ('ontic' constraints, von Glasersfeld, 1995, 52). The automatic interpretation of 'realism' in the correspondence sense is one of the myriad conceptual traps in this domain.

An evolutionary epistemological version of constructivism is required in order to account for obvious realities, such as the existence of representation and representational error. It is necessary, therefore, as a background set of assumptions, for exploration of any epistemological phenomena or endeavor. In particular, it is necessary for the understanding and planning of education.

FUNCTIONAL SCAFFOLDING AND SELF SCAFFOLDING

Functional scaffolding

I will illustrate some of the fecundity of a variation and selection evolutionary epistemology with a model that follows readily from it: a functional notion of scaffolding. This is related to scaffolding as commonly understood in contemporary literature, but constitutes a generalization of it (Bickhard 1992).

Constructions are retained only if they succeed – are 'viable'. This, of course, is relative to what selection pressures are actually being encountered by the organism – perhaps a child. If a particular task requires too complex a construction before such a construction will work, then the child is not likely to hit upon that complex construction, and is, therefore, not likely to master the task at all.

On the other hand, if some of the selection pressures normally involved in the task are 'artificially' blocked or removed, then some less complex constructions might succeed and be retained. If there is a trajectory of these protected – scaffolded – less complex constructions that results ultimately in the full successful construction, then provision of such scaffolding by a parent or teacher could make possible development or learning that would not otherwise be possible. Bruner's classic example is the great effort that parents put into the interpretations of communication attempts from infants that would otherwise not succeed in communicating (Bruner 1975).

Burner's notion of scaffolding (derived from Vygotsky's model of the zone of proximal development) turns on the provision of knowledge that the developing child does not otherwise have. Supplemented in this way, the child can succeed in tasks not otherwise possible, and can learn – internalize – the supplemental knowledge him- or herself, thus becoming competent to the task without the scaffolding.

Providing supplemental knowledge is one way in which selection pressures that would otherwise require that knowledge can be mitigated. But it is not the only way. Selection pressures can often be blocked without any knowledge of what would satisfy those selection pressures being provided, or being present at all. It is this broader functional view of scaffolding that the variation and selection model generates.

Functional scaffolding can be provided, for example, 'simply' by limiting the problems to which the child is exposed to simpler versions, at least at first. Protected environments, and curricular design are generally based on this intuition, but without the guide of a scaffolding model.

Self-scaffolding

Of greater importance, however, is that functional scaffolding, without the supplemental provision of otherwise absent knowledge, is something that the individual can do for him- or herself – and can *learn* to do for him- or herself – and can be *taught* to do for him- or herself. Such self-scaffolding is not a coherent notion within the standard view: a person cannot provide to himself supplemental knowledge that he does not have. Yet self-scaffolding, and the development of self-scaffolding skills, is arguably among the most important aspects of development and education alike (Bickhard 1992). Certainly it is an aspect that is not understandable without an evolutionary epistemology.

Examples of self-scaffolding would be: breaking a problem down into subproblems; moving to simpler ideal or analogous cases; looking for and learning to use resources that might not be necessary when the task is mastered; temporarily suspending constraints; and so on. It is obvious that these are important, even essential, skills and that they must be learned. Self-scaffolding offers a coherent model for such phenomena, an explanation for the importance of learning such skills, and a guide for the educational scaffolding of the learning of such self-scaffolding skills.

There is much more to be addressed concerning scaffolding and self-scaffolding (Bickhard 1992), but the intent here is more simply to illustrate the power of an evolutionary epistemological perspective. Functional and self-scaffolding are two examples among many of that modeling power (Bickhard 1992b). They show that an evolutionary epistemology, thus a constructivism, is not only necessary to avoid the traps of naive encodingist realism-idealism, but that it also offers useful new perspectives of its own. Note, in this regard, that the notion of functional scaffolding can not be formulated within an idealism. With idealism, there is nothing to impose selection pressures, and, therefore, there are no selection pressures to block or mitigate.

CONCLUSION

Constructivisms abound. Most of them are united in a recognition of the failure of classical correspondence-encoding models of representation. The positive proposals that they offer, however, vary wildly. In particular, idealist constructivisms, whether a solipsistic individual idealism, or a social-linguistic idealism, simply recapitulate the classic shift from realism to idealism, and thereby remain trapped in the correspondence-encoding framework. They do not transcend any of the aporia of that framework that they seek to escape. Instead, they create the additional serious threat of a nihilism and relativism.

Problems are created by the assumption that representations are a type of correspondence. Only when that view is rejected in favor of a pragmatic view, an account of the emergence of representation out of interaction, can these classic deadends be avoided. Such a pragmatic model of representation, in turn, *forces* an evolutionary epistemological constructivism. Such a constructivism is fundamentally different from an idealist constructivism (or an empiricist constructivism), and we mangle the issues if we confuse or misinterpret them.

REFERENCES

Bickhard, M.H.: 1988, 'Piaget on Variation and Selection Models: Structuralism, Logical Necessity, and Interactivism', *Human Development* **31**, 274–312.

Bickhard, M.H.: 1991, 'The Import of Fodor's Anticonstructivist Arguments', in L. Steffe (ed.), *Epistemological Foundations of Mathematical Experience*, Springer-Verlag, New York, pp. 14–25.

Bickhard, M.H.: 1992, 'Scaffolding and Self Scaffolding: Central Aspects of Development', in L.T. Winegar and J. Valsiner (eds.), *Children's Development within Social Context: Research and Methodology*, Erlbaum, Hillsdale, NJ, pp. 33–52.

Bickhard, M.H.: 1992b, 'How Does the Environment Affect the Person?', in L.T. Winegar and J. Valsiner (eds.), *Children's Development within Social Context: Research and Methodology*, Erlbaum, Hillsdale, NJ, pp. 63–92.

Bickhard, M.H.: 1993, 'Representational Content in Humans and Machines', *Journal of Experimental and Theoretical Artificial Intelligence* **5**, 285–333.

Bickhard, M.H.: 1993b, 'On Why Constructivism Does Not Yield Relativism', *Journal of Experimental and Theoretical Artificial Intelligence* **5**, 275–284.

Bickhard, M.H.: 1995, 'World Mirroring versus World Making: There's Gotta be a Better Way', in L. Steffe and J. Gale (eds.), *Constructivism in Education*, Erlbaum, Hillsdale, NJ, pp. 229–267.

Bickhard, M.H.: 1995b, 'Intrinsic Constraints on Language: Grammar and Hermeneutics', *Journal of Pragmatics* **23**, 541–554.

Bickhard, M.H.: in preparation, *The Whole Person: Toward a Naturalism of Persons*, Harvard University Press, Cambridge, MA.

Bickhard, M.H. & Campbell, R.L.: in press, 'Topologies of Learning and Development', *New Ideas in Psychology*.

Bickhard, M.H. & Terveen, L.: 1995, *Foundational Issues in Artificial Intelligence and Cognitive Science: Impasse and Solution*, Elsevier Scientific, Amsterdam.

Bruner, J.S.: 1975, 'The Ontogenesis of Speech Acts', *Journal of Child Language* **2**, 1–19.

Campbell, D.T.: 1974, 'Evolutionary Epistemology', in P. A. Schilpp (ed.), *The Philosophy of Karl Popper*, Open Court, LaSalle, IL, pp. 413-463.

Campbell, R.L. & Bickhard, M.H.: 1986, *Knowing Levels and Developmental Stages*, Karger, Basel.

Churchland, P.M.: 1989, *A Neurocomputational Perspective*, MIT Press, Cambridge, MA.

Fodor, J.A.: 1981, 'The Present Status of the Innateness Controversy', in J. Fodor, *RePresentations*, MIT Press, Cambridge, MA, pp. 257-316.

Fodor, J.A., Bever, T. & Garrett, M.: 1974, *The Psychology of Language*, McGraw-Hill, New York.

Gadamer, Hans-Georg: 1975, *Truth and Method*, Continuum, New York.

Gadamer, Hans-Georg: 1976, *Philosophical Hermeneutics*, University of California Press, Berkeley.

Gergen, K.J.: 1995, 'Social Construction and the Educational Process', in L. Steffe & J. Gale (eds.), *Constructivism in Education*, Erlbaum, Hillsdale, NJ, pp. 17-39.

Gier, N.F.: 1981, *Wittgenstein and Phenomenology*, SUNY Press, Albany.

Joas, H.: 1993, 'American Pragmatism and German Thought: A History of Misunderstandings', in H. Joas, *Pragmatism and Social Theory*, University of Chicago Press, Chicago, pp. 94-121.

Kuhn, D.: 1992, 'Cognitive Development', in M. H. Bornstein & M. E. Lamb (eds.), *Developmental Psychology: An Advanced Textbook*. Third Edition, Erlbaum, Hillsdale, NJ, pp. 211-272.

Lourenço, O. & Machado, A.: 1996, 'In Defense of Piaget's Theory: A Reply to 10 Common Criticisms', *Psychological Review* **103**(1), 143-164.

Rescher, N.: 1980, *Scepticism*, Rowman and Littlefield, Totowa, NJ.

Rorty, R.: 1979, *Philosophy and the Mirror of Nature*, Princeton University Press, Princeton, NJ.

Rorty, R.: 1987, 'Pragmatism and Philosophy', in K. Baynes, J. Bohman & T. McCarthy (eds.), *After Philosophy: End or Transformation*, MIT Press, Cambridge, MA, pp. 26-66.

Rosenthal, S.B.: 1983, 'Meaning as Habit: Some Systematic Implications of Peirce's Pragmatism', in E. Freeman (ed.), *The Relevance of Charles Peirce*, Monist, La Salle, IL, pp. 312-327.

von Glasersfeld, E.: 1979, 'Radical Constructivism and Piaget's Concept of Knowledge', in F. B. Murray (ed.), *The Impact of Piagetian Theory*, University Park Press, Baltimore, pp. 109-122.

von Glasersfeld, E.: 1981, 'The Concepts of Adaptation and Viability in a Radical Constructivist Theory of Knowledge', in I. E. Sigel, D. M. Brodzinsky & R. M. Golinkoff (eds.), *New Directions in Piagetian Theory and Practice*, Erlbaum, Hillsdale, NJ, pp. 87-95.

von Glasersfeld, E.: 1995, *Radical Constructivism*, Falmer Press.

Constructivisms and Objectivity: Disentangling Metaphysics from Pedagogy*

RICHARD E. GRANDY

Philosophy Department, Rice University MS 14, 6100 Main Street, Houston, TX 77005-1892, U.S.A.

ABSTRACT: We can distinguish the claims of cognitive constructivism from those of metaphysical constructivism, which is almost entirely irrelevant to science education. Cognitive constructivism has strong empirical support and indicates important directions for changing science instruction. It implies that teachers need to be cognizant of representational, motivational and epistemic dimensions which can restrict or promote student learning. The resulting set of tasks for a science teacher are considerably larger and more complex than on the older more traditional conception, but the resources of cognitive sciences and the history of science can provide important parts of the teachers intellectual tool kit. A critical part of this conception of science education is that students must develop the skills to participate in epistemic interchanges. They must be provided opportunities and materials to develop those skills and the classroom community must have the appropriate features of an objective epistemic community.

INTRODUCTION

This paper is an attempt to construct a philosophical background and underpinnings of the SEPIA project. SEPIA stands for Science Education Portfolio Instruction and Assessment, and was an NSF-funded research effort carried out in Pittsburgh area middle schools jointly by researchers from the University of Pittsburgh and ETS. Some of the applied pedagogy and implementation of the project can be found in Duschl & Feather (1995) In the best spirit of rational reconstruction, however, this is not a philosophical prologemenon which existed in articulated form before the project but rather emerged through discussion and modifications. Moreover, agreement in details is not to be expected among the various participants.

CONSTRUCTIVISMS: COGNITIVE, EPISTEMIC AND METAPHYSICAL

Since constructivisms are a dominant topic these years (Fosnot 1993; Giannetto 1992; Glasson 1992; Goldin 1990, Matthews & Davson-Galle 1992; Matthews 1994; O'Loughlin 1992, 1993; Von Glasersfeld 1990, 1992), and since SEPIA incorporates a number of constructivist principles, it will be helpful to delineate those principles and to distinguish them from some other principles which are not included. This may be a more generally useful exercise, of course, since it seems to me that there is not sufficient clarity about the variations on constructivism, let alone their relations and implications. I also believe that an understanding of the various elements and kinds of constructivism will be helpful in evaluating what is required of teachers in implementing SEPIA or any other curriculum that incorporates these important elements of constructivism.

There are a wide range of terms, and I am sure I will offend some authors by using the following distinctions rather than theirs. I distinguish cognitive constructivism, epistemic constructivism, and metaphysical constructivism.

Cognitive constructivism is the view that individual cognitive agents understand the world and make their way around in it by using mental representations that they have constructed. What they could in principle construct at a given time depends on the conceptual, linguistic, and other notational resources, e.g., mathematics and graphing, at their disposal and on their current representations of the world that they have constructed through their personal history. What they actually construct depends also on their motivations and on the resources of time and energy available to devote to this particular task.

By *metaphysical constructivism* I mean the (collection of) views that the furniture of the world is constructed by us. This view can be subdivided into the individualistic, which postulates individual constructions of individual worlds, and the social, which postulates social constructions of shared worlds. This view typically contrasts with metaphysical realism, the view that (much of) the furniture of the world exists independently of minds and thoughts. There are some obvious issues for the social sciences that I will not explore here, since social institutions are clearly human creations; the implications for geography or psychology or human biology are unclear and will also not be explored here.

Metaphysical realism itself comes in a range of positions on the optimism/pessimism scale with regard to the knowability of the structure. Optimistic metaphysical realism holds that not only does the universe have an intrinsic structure, that God used a blueprint if you like, but that the structure is knowable in principle by humans – we could understand the blueprint. Less optimistic, though still guardedly hopeful, versions would be that we can develop representations which are approximately correct descriptions of some aspects of the universe. How either of these positions is justified philosophically is a matter we will not linger over here, for my main point is that these issues are irrelevant for science education once we understand fully the implications of cognitive constructivism.

A metaphysical realist who accepts cognitive constructivism must recognize that whatever knowledge is attained or even attainable about the ultimate structure of the universe must be represented in the constructions of the cognizer. While accepting cognitive constructivism has very important consequences for the teacher, which I will elaborate on shortly, once you have embraced cognitive constructivism it makes very little difference what attitude one has toward the metaphysical realist issues. However independent of us the structure of the universe may be, what we can achieve by way of producing more knowledgeable students depends on the representations they can construct. This seems to me of great importance because if Constructivism is presented as a package which includes both cognitive constructivism and metaphysical anti-realism then teachers who have long-standing philosophical inclinations toward realism will find the package unpalatable. Cognitive constructivism is a relatively empirical theory which has strong evidential support from psychology, artificial intelligence and education; metaphysical realism is a venerable philosophical doctrine supported by philosophical arguments, and subject to equally venerable philosophical objections.

Accepting cognitive constructivism has very significant consequences for understanding the tasks, and the demands of the tasks, required of the science teacher (Bloom 1992). All science teachers have, and must necessarily always have had, a philosophy of science – a set of beliefs about the nature of scientific inquiry, of scientific progress, of scientific reasoning, of scientific data, theories, and so on. Often this has been at least somewhat unconscious and implicit, often acquired unreflectively along with the content knowledge in science classes. And often, in the past at least, this philosophy of science incorporated beliefs in the continuous linear progress of science, of the empiricist inductive scientific method, in the immutability of scientific facts, in scientific realism, perhaps even metaphysical realism, and so on. It has often included the philosophy of science education which is described as direct teaching (Duschl 1990), or, as I think of it, the modified Dragnet theory of teaching. Unlike the old Dragnet show we don't just give them the facts, but on that model we do just give them the facts, definitions and theories, and nothing but the facts, definitions and theories.

Whatever the remainder of one's philosophy of science, to accept cognitive constructivism means recognizing that each student constructs a representation based on their experience, including but by no means limited to teachers verbal input. The teacher must assess the extent to which the student's representation is isomorphic to the teacher's, but of course cognitive constructivism applies reflexively and the teachers have no direct infallible access to the students' representations but instead construct their own representations of the students' representations. Since one of the typical student's motivations, for better or worse, is to please the teacher it may also be valuable for the teacher to construct a representation of the student's representation of the teacher's representation.

The practical issues of this process can be discussed under the heading of assessment Duschl and Feather (1995), but I want to make some general more philosophical points about the process. Teachers are necessarily pursuing this process under time constraints and must repeatedly balance the potential value of further exploring the students representation in all of its detailed uniqueness against categorizing the students representation as sufficiently similar to others seen in the past to allow a particular course of further instruction to be developed without further investigation.

Of course that is only part of the task, for the teacher may well need to understand also why the student has constructed that particular representation. The divergence from the desired kind of representation may result from lack of the tools to construct an alternative, lack of accepted evidence that the current representation is insufficient or lack of motivation to construct an alternative (Ames 1990). The next step to bringing about desired change will likely depend on which of these factors is prevalent, and this implies that the teacher must have an understanding of motivational psychology, of the evidence the student accepts, what the student counts as evidence, and on what conceptual tools the student can make use of. If this is correct, the the conceptual change movement was in the right direction, but the process of instigating conceptual change in the student is more complicated than was probably initially recognized.

OBJECTIVITY AND CHANGE

Some science educators are unwilling to pronounce any student representation a 'misconception'. The reasons for this reluctance are important to analyze and understand. There is a very important positive aspect to this taboo which stems from the insight that the students produce the best representations that they are capable of producing at the time given their information, conceptual and motivational constraints. The student is not to be faulted. But if we are unwilling to evaluate representations, unwilling to judge some representations and understandings as more accurate, more general, more consistent than others, then there is no reason to teach 'science'. Why spend our time on such a frustrating activity if we do not think that the student is in principle capable of a representation which is in some important sense an improvement? If the representations of the teachers or of the scientific community are not in some judgmental way better than those of neophyte students then there is nothing to teach and time would be better spent on spelling.

Having said that, and emphasizing the difference between judging the representation and judging the student, we should note that the fact that the teacher sees room for improvement does not mean that the best way to proceed is by directly criticizing the students representation. One important aspect of the cognitive constructivist view is that the students repertoire typically includes more than one way of representing a situation. Just as we can draw various maps representing various aspects of the world – highways, rainfall, elevation, population – the cognitive agent represents any situation in sundry ways. We recognize now that it is not just a matter of enabling the student to construct a Newtonian representation of a situation, but that there is the further project of ensuring that the Newtonian representation, rather than the intuitive physics representation, is applied. There is nothing wrong with having multiple theories or representations – the best-known examples are the use of Newtonian physics for large slow objects, and relativity or quantum mechanics for the very fast or very small. This use of multiple representations is probably ubiquitous in the sciences – Cartwright (1983) cites a number of such cases, including six competing mathematical treatments of quantum damping (p. 78 ff) as part of her argument for the ontological priority of causes over laws. But there is an important and somewhat elusive element of expertise in knowing what representation to deploy. And, most importantly, there is abundant evidence that students continue to deploy intuitive versions of Aristotelian physics after training at the college level in Newtonian mechanics (Bruer 1993, 130 ff).

It will be helpful to distinguish, using Megill's terms, absolute from disciplinary objectivity. Absolute objectivity was the goal of some of the mathematical explorations of inductive logic during the heyday of logical positivism, and has long been a philosophical Holy Grail. Like metaphysical realism, it is irrelevant, in my view, to the process of science pedagogy since absolute objectivity is at best an ideal and in the classroom we are but beginning the process of developing a sense of objectivity. Absolute objectivity requires criteria of validity which are invariant over time or culture or discipline. In contrast, it seems to me that a very worthwhile, and manageable goal, is to develop and nurture disciplinary objectivity.

> Disciplinary objectivity emphasizes not universal criteria of validity but particular, yet still authoritative, disciplinary criteria. It emphasizes not the eventual convergence of all inquirers of good will but the proximate convergence of accredited inquirers within a given field (Megill 1991, p. 305).

Megill qualifies his adjective 'disciplinary' with a footnote saying that many of the criteria tend to be even more specific than disciplinary, i.e., originating from subdisciplines. This is important for the classroom for the culture of the classroom, the discipline which we can hope that students will construct is not professional biology or geology, but an age appropriate variant of that, which is also limited by the constraints on time, equipment and other resources. The goal is to create in the classroom a subculture which is in some appropriate ways related to the discipline under study.

EPISTEMIC CONSTRUCTIVISM: SOCIAL OR INDIVIDUAL?

In my presentation of constructivisms and their relations in the first section I ignored *epistemic constructivism*, the view that knowledge is constructed by us rather than directly imbibed from the environment. This seems to me a reasonable consequence of cognitive constructionism, though a thorough defense would require a detailed discussion of the various possible analyses of knowledge. Rather than engage in that enterprise, I would like to flag an important aspect of the use of the term 'construction' that has, I think, been insufficiently remarked on. The use of the term 'constructivism' is arguably a metaphor extending to the abstract a notion that makes good physical sense – we construct dams, buildings, airplanes.

What I want to note is that in the case of physical construction there are always important constraints on the construction process if we are constructing for some useful purpose and not simply to expend time and energy. Some methods of construction are more efficient than others, some are faster, some are slower and more expensive but produce a more enduring product. Many shoddy methods of construction produce nothing of value. Surprising little discussion has been expended on the issue of the methods and materials of construction that go into constructing representations of scientific objects, data and theories.

Accepting that knowledge is constructed, a natural metaphysical question that arises at this point is whether one sees the scientific group or the individual as the basic unit of analysis and explanation of knowledge. Two fairly representative but divergent positions are the following which are, or were, held by two distinguished philosophers of science from the University of Minnesota:

> What I propose... is a much more thorough going contextualism than the one which urges us to remember that scientific inquiry occurs in a social context, or even that scientists are social actors whose interests drive their scientific work. What I urge is a contextualism which understands the cognitive processes of scientific inquiry not as opposed to the social, but as themselves social. This means that normativity, if it is possible at all, must be imposed on social processes and interactions, that is, that the rules or norms of justification that distinguish knowledge (or justified hypothesis-acceptance) from opinion must operate at the level of social as opposed to individual cognitive processes (Longino 1992).

In contrast

> The conclusion is simple. The most promising approach to a general theory of science is one that takes individual scientists as the basic units of analysis. It follows that we must look to the cognitive sciences for our most basic models, for it is these sciences that currently produce the best causal models of the cognitive activities of individual human agents (Giere 1989).

The view I am advocating accepts neither model, but sees the continuing dynamic interaction between group and individuals as critical. (Cf. Cobb 1994a,b; and Driver 1994, for related arguments.) '/although learning science involves social interactions, ... we have argued that individuals have to make personal sense of newly introduced ways of viewing the world' (Driver 1994, p. 11). Objective knowledge is the result of an interactive process between individuals and community. It is essential to see that although a group is in an obvious sense constituted at a given time by a set of individuals, as a group changes over time members are attracted to the group or become part of it because of the properties of the group as a whole. The group, and the perception of the group, shape the cognitive behavior of those who join it. Moreover, epistemic evaluation seems appropriate for both individual and group processes, although the units and the measure of evaluation differ.

At the professional level of science education, there is a tension between ensuring that everyone who is given the formal credentials of the discipline share the fundamental values and concepts so that the coherence of the discipline is preserved, and ensuring that innovative thinkers who may question even fundamental values and concepts are not excluded so that the possibility of creative innovation is preserved. In other words, both the group, as a group, and the individuals who constitute it must have appropriate characteristics in order for there to be a significant cognitive activity worth calling knowledge. One analysis of the role of the group is that of Longino (1994).

She lists four conditions for a community to meet in order for a consensus to qualify as knowledge:

1. There must be publicly recognized forums for the criticism of evidence, of methods, and assumptions and reasoning.
2. There must be uptake of criticism. The community must not merely tolerate dissent; its beliefs and theories must change over time in response to the critical discourse taking place within it.
3. There must be publicly recognized standards by reference to which theories, hypotheses, and observational practices are evaluated and by appeal to which criticism is made relevant to the goals of the inquiring community...
4. Finally, communities must be characterized by equality of intellectual authority. What consensus exists must be the result not of the exercise of political or economic power, or of the exclusion of dissenting perspectives, but a result of critical dialogue in which all relevant perspectives are represented. This criterion is meant to impose duties of inclusion; it does not require that each individual, no matter what her or his past record or state of training, should be granted equal authority... (Longino 1994, pp. 144–5).

These requirements were written with the professional community, not the classroom in mind, but I think they are a reasonable set of guidelines for the classroom scientific community as well. There are a number of crucial and vague terms, but while these can be fleshed out in somewhat more detail, the exact details will have to be developed and negotiated separately in each individual case.

INDIVIDUAL, SOCIETY AND HISTORY

For either individual or social construction of science more is needed than theories, data and instruments. What is missing are the epistemic connections that relate theories to supporting data, to conflicting theories, to anomalous data, to equivocal data. The concept of a data domain, as developed by Ackermann (1985) and of the importance of anomalous data (Chinn & Brewer 1993) within that domain needs to be emphasized. While the individual has some freedom to argue against the grain of the scientific community, to a large extent what can be taken as data and what is disqualified, what is strong evidence and what is weak evidence, is always judged against the background provided by the community's experience with the theories, the data domain and the instruments in question. The data domain may be very refined, as in the case of professional level well-established sciences, or much more in flux, as it will be in the classroom scientific community, but the demarcation between what counts and what does not, however fluctuating it may be over time, is critical to the ongoing enterprise. And initiation into the process of constructing data, evaluating data, citing data and contesting data are all part of the individual's skills in the social setting.

The role of history of science in science teaching has received much attention (e.g, Aikenhead 1992; Gil-Perez 1992; Niedderer 1992; Matthews 1994, and further references there) and I cannot resist adding a few sentences on how the conception above relates to the use of history. The history of the development of a particular scientific theory or conception – the Copernican system, plate tectonics, Darwinian evolution – is one or more routes by which reasonable inquirers arrived at a conclusion. The starting point for late-twentieth century students is not the same as for the historical inquirers. The most obvious example being that most students 'know' that the earth goes around the sun when they come to science class, even though they are frequently unable to develop the appropriate conclusions from that knowledge. The reasons are complex, but they include in most cases the fact that they bring to bear a version of intuitive physics. This latter bears many resemblances to the sophisticated neo-Aristotelian physics of the sixteenth century, but it would be a mistake to treat them as identical. And the motivations of sixteenth and seventeenth century intellectuals were different in many very important respects than those of our current students.

On my view, knowing the history of a scientific development provides the teacher with a set of arguments and experiments and an epistemic route from one cognitive locus to another. This is often an important part of the tool kit that can be used to assist learners in constructing their own representations more satisfactorily. But as I have outlined above, the history of science by itself is far from sufficient for the teacher confronted with a very complex set of tasks. I see the extent of the utility of history of science as being subject to possible empirical study. A second point is that independent of issues of bringing the students constructions to a different state by calling on the same arguments and experiments that were historically used, reflection on the history of science itself can provide important fodder for the epistemic learning mill. The concepts of data, anomalous data, questionable data and so on can be illustrated in the history of science as well as in the classroom productions. This is another step in forging the cultural links between the classroom inquiry and the larger scientific process.

CONCLUSION

I have argued that we can distinguish the claims of cognitive constructivism from those of metaphysical constructivism. Cognitive constructivism has strong empirical support and indicates some important directions for changing science instruction. It implies that teachers need to be cognizant of representational, motivational and epistemic dimensions which can restrict or promote student learning. Metaphysical issues are irrelevant to the pedagogical enterprise except when explicit philosophical issues arise. The resulting set of tasks for a science teacher are considerably larger and more complex than the older more traditional conception, but the resources of cognitive sciences and the history of science can provide important parts of the teachers intellectual tool kit.

A critical part of this conception of science education as informed by cognitive constructivism is that the students must develop the skills to participate in the epistemic interchanges that take place in scientific communities. They must be provided opportunities and materials to develop those skills and the classroom community must have the appropriate features of an objective epistemic community.

NOTE

* Some of the funding for this research was provided by a grant from the National Science Foundation (MDR-9055574) to the University of Pittsburgh and the Educational Testing Service. The opinions expressed do not necessarily reflect the positions or policies of NSF and no official endorsement should be inferred. The paper was initially presented at the Third International, History, Philosophy and Science Teaching Conference, Minneapolis 1995.

REFERENCES

Ackermann, R.J.: 1985, *Data, Instruments, and Theory: A Dialectical Approach to Understanding Science*, Princeton University Press, Princeton.

Aikenhead, G.: 1992, 'How to Teach the Epistemology and Sociology of Science in a Historical Context', in S. Hills (ed.), *Second International HPS&ST Proceedings*, University of Kingston, Kingston, pp. 23–34.

Ames, C.A.: 1990, 'Motivation: What Teachers Need to Know', *Teachers College Record* **91**, 409–421.

Bloom, J.: 1992, 'Contextual Flexibility: Learning and Change from Cognitive, Sociocultural, and Physical Context Perspectives', in S. Hills (ed.), *Second International HPS&ST Proceedings*, University of Kingston, Kingston, pp. 115–126.

Bruer, J.T.: 1993, *Schools for Thought: A Science of Learning in the Classroom*, MIT Press, Cambridge.

Cartwright, N.: 1983, *How the Laws of Physics Lie*, Clarendon Press, Oxford.

Chinn, C. & Brewer, W.: 1993, 'The Role of Anomalous Data in Knowledge Acquisition:

A Theoretical Framework and Implicatioins for Science Instruction', *Review of Ed Research* **63**, 1–50.

Cobb, P.: 1994a, 'Constructivism in Mathematics and Science Education', *Educational Researcher* **23**, 4.

Cobb, P.: 1994b, 'Where Is the Mind? Constructivist and Sociocultural Perspectives on Mathematical Development', *Educational Researcher* **23**, 13–23.

Driver, R. et al.: 1994, 'Constructing Scientific Knowledge in the Classroom', *Educational Researcher* **23**, 5–12.

Duschl, R.A. & Feather, R.: 1995, 'Developing and Nurturing Objectivity in Science Classrooms', in *Proceedings of the Third International Conference on History and Philosophy of Science and Science Teaching*, University of Minnesota Press, Minneapolis, MI, pp. 314–325, Vol. 1.

Duschl, R.A.: 1990, *Restructuring Science Education: The Importance of Theories and Their Development*, Columbia University Press, New York.

Fosnot, C.: 1993, 'Rethinking Science Education: A Defense of Piagetian Constructivism', *Journal of Research in Science Teaching*, 30(9), 1189–1201.

Giannetto, E.: 1992, 'The Relations Between Epistemology, History of Science and Science Teaching from the Point of View of the Research on Mental Representations', in S. Hills (ed.), *Second International HPS&ST Proceedings*. University of Kingston, Kingston, pp. 359–374.

Giere, R.: 1989, 'The Units of Analysis in Science Studies', in S. Fuller et al. (eds.), *The Cognitive Turn*, Kluwer Academic Publishers, Dordrecht, pp. 3–11.

Giere, R.: 1988, *Explaining Science: A Cognitive Approach*, University of Chicago Press, Chicago, IL.

Gil-Perez, D.: 1992, 'Approaching Pupil's Learning to Scientific Construction of Knowledge: Some Implications of the History and Philosophy of Science in Science Teaching', in S. Hills (ed.), *Second International HPS&ST Proceedings*, University of Kingston, Kingston, pp. 375–390.

Glasson, G. et al.: 1992, 'Social Constructivism in Science Learning: Toward a Mind-World Synthesis', in S. Hills (ed.), *Second International HPS&ST Proceedings*, University of Kingston, Kingston, pp. 399–406.

Goldin, G.: 1990, 'Epistemology, Constructivism, and Discovery Learning in Mathematics', in R.B. Davis, C. Maher & N. Noddings (eds.), *Constructivist Views on The Teaching and Learning of Mathematics*, NCTM, pp. 31–47.

Longino, H.: 1992, 'Essential Tensions – Phase Two: Feminist, Philosophical and Social Studies of Science', in E. McMullin (ed.), *The Social Dimensions of Science*, U Notre Dame Press, Notre Dame, IN, pp. 198–216.

Longino, H.: 1994, 'The Fate of Knowledge in Social Theories of Science', in F.F. Schmitt (ed.), *Socializing Epistemology: The Social Dimensions of Knowledge*, Rowman & Littlefield, Lanham, Maryland, pp. 135–157.

Matthews, M. & Davson-Galle, P.: 1992, 'Constructivism and Science Education: Some Cautions and Comments', in S. Hills (ed.), *Second International HPS&ST Proceedings*, University of Kingston, Kingston, pp. 135–144.

Matthews, M.: 1994, *Science Teaching: The Role of History and Philosophy of Science*, Routledge, New York.

Megill, A.: 1991, 'Four Senses of Objectivity', *Annals of Scholarship* **8**, 301–320.

Niedderer, H.: 1992, 'Science Philosophy, Science History and the Teaching of Physics', in S. Hills (ed.), *Second International HPS&ST Proceedings*, University of Kingston, Kingston, pp. 201–214.

O'Loughlin, M.: 1993, 'Some Further Questions for Piagetian Constructivists: A Reply to Fosnot', *Journal of Research in Science Teaching* **30**(9), 1203–1207.

O'Loughlin, M.: 1992, 'Rethinking Science Education: Beyond Piagetian Constructivism Toward a Sociocultural Model of Teaching and Learning', *Journal of Research in Science Teaching* **29**(8), 791–820.

Von Glasersfeld, E.: 1992, 'A Constructivist Approach to Experimental Foundations of

Mathematical Concepts', in S. Hills (ed.), *Second International HPS&ST Proceedings*, University of Kingston, Kingston, pp. 553–572.

Von Glasersfeld, E.: 1990, 'An Exposition of Constructivism: Why Some Like it Radical', in R.B. Davis, C. Maher & N. Noddings (eds.), *Constructivist Views on The Teaching and Learning of Mathematics*, NCTM, Reston VA, pp. 19–30.

Social Constructivism, the Gospel of Science, and the Teaching of Physics

HELGE KRAGH

History of Science Department, Aarhus University, Ny Munkegade, 8000 Aarhus, Denmark
E-mail: ievhhk@dfi.au.dk

ABSTRACT. During the last two decades, science studies have increasingly been dominated by ideas related to social constructivism and the sociology of scientific knowledge. This paper offers a critical examination of some of the basic claims of this branch of science studies and argues that social constructivists cannot explain some of the most characteristic features of the physical sciences. The implications of social constructivism for science education are considered. I conclude that if education in physics consistently followed the philosophy of sociology of scientific knowledge in its more extreme versions it would mean the end of physics. However, the rejection of social constructivism does not imply a rejection of social or cultural studies of science or their value in science education.

INTRODUCTION

In spite of all attempts to soften the divide between the scientific and the literary-humanistic fields of learning in our education systems the two branches are today as separate as ever. Of course the separation is not particular to the education system, where it merely appears as a manifestation of the more general culture gap that characterizes Western societies at large. Moreover, the relation between the two cultures is highly asymmetric, in higher education as well as in the larger society. Generally speaking, our public culture is predominantly literary-humanistic, a fact that is illustrated with clarity by comparing the mass media's, and especially the newspapers', treatment of, say, science and literature. In higher education the imbalance is mercilessly reflected in the students' choice of subjects, always a good indicator of the popularity of academic fields. Areas such as philosophy, literature, theology, anthropology and cultural sociology are booming in North America and many European countries where they attract large numbers of students. Physics, mathematics, chemistry and the engineering sciences, on the other hand, have increasing difficulties in recruiting enough students to their expensive schools and laboratories; and this is spite of massive efforts to attract students to these fields, supposed to be much more useful to modern society.

It is not the job of physicists and physics teachers to participate in this propaganda, but neither should they rest content with a situation where a growing number of students can graduate without knowing a thing about science and technology; or, even worse, where they graduate with the belief that they know what science is, based on the quasi-knowledge offered by fashionable philosophers, sociologists and literary critics. In the case of the United States, the public scientific illiteracy is illustrated by surveys that show that only 13 per cent of the adult population have a minimum understanding of the process of science, and 40 per cent are willing to grant astrology scientific status; 12 per cent of the physics teachers meet the minimum standards for course preparation work at the high school level and only 20 per cent of the high school graduates have taken physics courses of any kind (Holton 1993, p. 148).

Few would deny that science has been, and continues to be, a powerful economic and cultural force that has thoroughly shaped our material and spiritual situation. It would not be an exaggeration to claim that science is the single most important factor in the more recent history of humankind. It is therefore deeply paradoxical that our culture, which is very much the product of science and technology, is predominantly literary and humanistic and that science is still considered either a mysterious field of semi-gods like Richard Feynman and Stephen Hawking, or an instrumentalistic endeavour which is hostile to true culture. Now the last couple of decades have witnessed a booming interest in an attempt to demystify science and present it as deeply connected with human and social affairs. Is this trend – I am referring to the epistemic-sociological turn in science studies – a solution to the problems that face science in the areas of culture and education? At least if viewed from the perspective of most scientists and teachers of science, the answer is undoubtedly that not only do the new science studies offer no solution, they are positively harmful and essentially anti-scientific.

SCIENCE UNDER ATTACK

Ever since its take-off in the seventeenth century, science has continually been attacked and alternative forms of understanding nature have been proposed as substitutes for the scientific world view (Toulmin 1973; Dessaur et al. 1975; Nowotny and Rose 1979). According to one tradition of anti-science, the romantic or utopian tradition, science embodies cold rationality and violates nature by interfering with her – the scientist being likened to a rapist. The romantic tradition emphasizes subjectively and intuitively gained insight and opposes the canons of objectivity characteristic of ordinary science. Whereas the romantic opposition is anti-science and aspires to establishing a radically different view of nature, pseudo-sciences (such as astrology and parapsychology) are not against science in the same manner, but accuse established science of neglecting or dismissing a wide range of phenomena that cannot be explained according to the orthodox theories of science. The pseudo-scientist attacks scientific monopoly rather than science *per se* and argues for the right to other kinds of knowledge than the one defined officially. Still other kinds of counter-movements, sometimes referred to as critical or alternative science, focus on the political consequences of science. Rather than regarding science as a liberating force, a Prometheus, they see it as a modern Frankenstein monster which is opposed to true democratic and human development.

As indicated, the anti-science movement is extremely heterogeneous. It forms a wide spectrum including, for example, conservative religiously-based criticism as well as Marxist-oriented attempts to redefine science. Vitalism, anti-intellectualism and versions of *Lebensphilosophie* form one pool which historically has followed the development of modern science and is associated with the romantic opposition. Of more interest is the more recent philosophically and sociologically based science criticism that emerged in the 1960s, in part inspired by Kuhn's influential revolt against positivism and critical rationalism. Although Kuhn's views cannot be interpreted in favour of anti-science, Paul Feyerabend's vigorous criticism can reasonably be seen as belonging to the anti-science tradition. According to Feyerabend, science is purely ideological and executes a mental dictatorship on line with the one of the church in the Middle Ages. He therefore called for an abolition of obligatory science in schools and for a stop of government support to all science activities. Coming from a very different corner, a somewhat similar kind of anti-science is represented by left-wing philosophers and sociologists associated with the so-called Frankfurt school. According to Marcuse and Habermas, science is manipulative 'instrumental reason' and by its very nature opposed to political consciousness and human liberation. The exploitation, violence and social manipulation associated with modern science are not merely the results of wrong applications of science, but are claimed to be elements inherent in science itself.

These brief remarks only sketch a few of the anti-science or counter-culture positions that characterized the debate in the 1970s. Although noisy, they had relatively little influence on academic life and attempts to establish alternative forms of science – such as a new physics based on Tao philosophy or other forms of Eastern wisdom – failed to deliver what they promised. But at the same time as the anti-science movement declined an academically more respectable variant of science criticism saw the light of day, the sociology of scientific knowledge tradition associated with the Edinburgh school's strong programme. The works of the Edinburgh and Bath sociologists marked the start of an epistemic turn in sociology of science which since then has continued and developed into modern social constructivism.

Since this epistemic turn owed much to Kuhn's theory of science, and Kuhnian notions are still among the ammunition used by relativist historians and sociologists, it should be emphasized that Kuhn was never happy about the ways his theory was used. In fact, he came to disagree strongly with the social constructivist trend. Thus, in 1992 he explicitly criticized the view that 'what passes for scientific knowledge becomes, then, simply the belief of the winners'. Kuhn stated his position as follows: 'I am among those who have found the claims of the strong program absurd: an example of deconstruction gone mad. And the more qualified sociological and historical formulations that currently strive to replace it are, in my view, scarcely more satisfactory' (Kuhn 1992, p. 9).

Although constructivist sociologists deny that they are against science, and although there are significant differences between their views and those of, say, Feyerabend or Marcuse, there are also similarities. In particular, by regarding science as a social construction they deny that the scientific world view is grounded in nature and should therefore be given higher priority than any other world view. In practice, if not in theory, modern social constructivism has contributed to a revival of anti-science sentiments and a renewed polemics about the role of science in society and education. Some scientists have reacted strongly against the new form of 'higher superstition' and in general the relations between scientists and science analysts have hardened. The debate that followed the publication of Paul Gross and Norman Levitt's *Higher Superstition* (1994) – a brilliant, provocative and caustic attack on postmodernist science studies – has shown that at least some scientists have had enough of relativist sociologists' self-confident conceptions of science.

The physicist Steven Weinberg and the embryologist Lewis Wolpert are among those who have dismissed the new brand of science studies (Weinberg 1992, Wolpert 1992). The result has been an almost insurmountable barrier between the scientists and those who study science from a social and cultural perspective, unfortunately also threatening the relationship between science and more conventional forms of history and philosophy of science. As pointed out by the chemist Jay Labinger, not only are scientists in practice excluded from the purportedly interdisciplinary science studies programmes, but potential collaborators are frightened away by the extreme relativism and anti-science attitudes of many science studies scholars. As Labinger puts it, 'Trying to convince scientists to do something based on the premise that they are all wrong is not likely to be very successful' (Labinger 1995, p. 301).

SCIENCE AS A SOCIAL CONSTRUCTION

Social or cultural constructivist science studies exist in numerous variants, some more radical than others. There are even those which are compatible with empiricism and realism (Sismondo 1993). For the sake of brevity I ignore the differences between social, cultural, epistemic and other versions of constructivism and merely refer to 'constructivism' (which should not, of course, be confused with the didactical method of the same name). I shall here be concerned only with those versions which, in one way or other, consider science to be socially (or culturally) constructed in the strong sense that 'scientific knowledge originates in the social world rather than the natural world' (Woolgar 1983, p. 244). Insofar as constructivists acknowledge the existence of a natural world independent of social groups and mechanisms they regard it as merely a constraint that mildly influences the scientists' accounts, if at all. According to Harry Collins, an important school in the social studies of science embraces 'an explicit relativism in which the natural world has a small or non-existent role in the construction of scientific knowledge' (Collins 1981, p. 3).

The epistemology characteristic of constructivists is either relativistic or agnostic, in the sense that they do not admit any distinction between true and false accounts of nature (for which reason 'true' and 'false' invariably appear in citation marks in constructivist literature). Denying the existence of an objective nature, or declaring it without interest, scientists' accounts are all there is, and it is with these accounts the constructivist sociologist is solely concerned. How, then, do scientists manage to produce their results and build up a corpus of consensual knowledge about what they call nature? Not by wrestling with nature's secrets either in the laboratory or at the writing desk, but by negotiations, political decisions, rhetorical tricks, and social power. Since truth and logic are always relative to a given local framework, scientists' beliefs about nature are not inherently superior to those of any other group. As Andrew Pickering has concluded in his book about the social construction of the quark concept: 'The world of HEP [high energy physics] was *socially* produced. ... there is no obligation upon anyone framing a view of the world to take account of what twentieth-century science has to say' (Pickering 1984, pp. 406, 413).

The constructivist view of science is clearly at odds with the conventional view favoured by almost all scientists and science educators, namely, that science is a way of understanding increasingly more about nature by following certain rules, techniques and methods that are collectively known under the somewhat misleading term 'the scientific method'. The excesses of social constructivism have been severely criticized by scientists, philosophers and non-constructivist sociologists who have argued that the constructivist project is absurd, inconsistent, or empirically unfounded (e.g., Murphy 1994; Laudan 1990; Cole 1992).

There is, for example, the problem of reflexivity. If there are no context-independent truths, why should we believe that the constructivist account of science is better or more true than the conventional account? After all, the constructivist claims that his arguments are superior to those of the realist and not merely superior within the constructivist framework (in which case the claim would be tautological). Moreover, it is not clear at all what constructivists mean by the term 'social', which they often use in a way that would ordinarily include cognitive factors. Is it reasonable to say that the views of a scientist are governed by social interests when he defends a theory he believes is true? In that case all decisions in science would be social just because science takes place in a social context and the whole argument would tend to be tautological and hence empty. Peter Galison, among others, has argued against the constructivist conflation of 'social' and 'cognitive' and has shown how episodes in modern physics do not fit the constructivist programme (Galison 1987, pp. 258–9).

However, these and other objections seem to have had no major effect. Constructivism continues to be a popular approach among social and humanist scholars, although with signs of growing insulation and secterianism. Rather than addressing some of the more general weaknesses in the constructivist programme, I want to mention a few specific points that this approach to science studies seems unable to explain in a satisfactory way.

(i) *Discoveries*. Physicists sometimes happen to discover phenomena that are absolutely unexpected. Examples are Röntgen's discovery of his rays in 1895 and Kammerlingh-Onnes's discovery of superconductivity in 1911. In other cases the phenomena were predicted by theory, but the experimentalists were in fact unaware of the predictions, such as was the case with Anderson's discovery of the positron in 1932 and Penzias and Wilson's observation of the cosmic microwave radiation in 1965. Such discoveries and their subsequent histories are accounted for in a natural way according to the conventional view – that the objects or phenomena exist in nature or can be produced in the laboratory – but they seem inexplicable according to the constructivist theory. Yet another type of discovery that strongly indicates that knowledge in physics reflects how nature works, and not solely how scientists negotiate, is the quantitatively precise and confirmed predictions. The famous 1846 discovery of Neptune, guided by the predictions of celestial mechanics, is a case in point and so is the determination of the electron's anomalous magnetic moment in the late 1940s. From the point of view of constructivism, all this must appear either a mystery, a coincidence, or the result of conspiracy.

To be fair, constructivists have dealt extensively with discoveries. They argue, for example, that 'the validity of new discoveries should be examined as a sort of convention or belief' and that 'discoveries occur because they are made to occur socially by processes of social recognition' (Brannigan 1981, pp. 78, 169). The attribution of a discovery claim as a discovery is of course a social process, but the discovered objects and phenomena are surely parts of nature and cannot be negotiated away if the scientists should so decide. A related problem appears when constructivists are confronted with multiple discoveries, that is, discoveries made simultaneously and independently by different scientists. If the content of a scientific discovery is determined by specific, but contingent social circumstances, how is it that scientists working in very different contexts can make the same, or almost the same, discovery?

(ii) *Progress*. The progress of scientific knowledge is at the same time one of the most characteristic features of science and one of the most controversial. The subject can be discussed endlessly, but can it be seriously contested that physicists in the 1990s know more about electrical currents than Faraday did in the 1830s? Philosophical niceties apart, did Boyle know as much about chemical reactions as we do? It should be uncontroversial that we can accurately predict and account for vastly more about nature than our ancestors could two or three hundred years ago, and in this simple sense progress must be considered a fact. All the same, 'progressivism' has a bad reputation among constructivists, who either ignore or deny scientific progress. And no wonder, for roughly continual progress over many generations does not fit with constructivist ideas. Within this framework there is no explanation of scientific progress.

(iii) *Fallibility and corrigibility*. As stressed by Allan Franklin, among others, theories of physics are not only fallible, but also corrigible. It is a matter of fact that theories are routinely proved wrong by experiments and that scientists defending a theory often accept experimental data which contradict their theory. Scientists often feel forced to abandon a theory in the light of experimental evidence, in spite of having a vested interest in the theory. They do not normally cling to orthodox theories or suppress data disagreeing with these. As Franklin has remarked: 'Among the most important kinds of experiment are those that refute a well-confirmed theory or those that confirm an implausible theory. It is an experimenter's hope to find such unexpected results' (Franklin 1990, p. 137). Such behaviour seems incompatible with constructivists' notion of experiment, according to which an experiment can only be successful if it confirms the expected.

Collins and Shapin, in a discussion of the role of social studies of knowledge in physics education, thus claim that 'an experimental run which produces anomalous results is accounted, and assessed as, a failed run. This retrospective assessment procedure is one of the ways that consensus is maintained in normal science: anomalous results are excluded by fiat' (Collins and Shapin 1983, p. 286). The claim is simply contradicted by the history of physics, such as shown by the discovery of parity nonconservation in 1957 and numerous other experiments.

(iv) *Archaic evidence*. Modern scientists routinely rely upon historically remote observations and count these as correct even if they were made and interpreted in the context of a totally false theory. In astronomy and geophysics, in particular, such archaic or ancestral evidence is highly important. For example, ancient Babylonian observations of lunar eclipses have been used with success to determine the change in speed of the earth's rotation and this has again led to improved geophysical models of the interior of the earth (Stephenson 1982). From a realist point of view it makes sense to rely upon ancient observations, for these were about the same world as ours; the theories have changed, but the objects of the theories (such as the earth and the moon) remain the same. However, this kind of use of old observations must appear mysterious and irrational within a constructivist perspective (Trout 1994). According to the consistent constructivist, a theory does not describe the world, but constitute it in the sense that the subject matter of science is constructed by the background theoretical assumptions of the local scientific community. Archaic evidence such as Babylonian observations therefore cannot play any confirmatory role in contemporary science. And yet they do.

(v) *Stable experimental results*. Social constructivists exploit to the extreme the fact that in many situations of experimental science the experimentalists disagree. However, in most or all cases this can be explained without recourse to sociological mechanisms. More to the point, the emphasis on disagreement in experiments is grossly exaggerated. Without exact documentation, I wil claim that most experiments in physics do not give rise to controversies or major disagreements of the kind known from cold fusion, gravitational wave research or the Millikan-Ehrenhaft dispute concerning the existence of subelectrons. In most experiments the data quickly stabilize and consensus is achieved in an undramatic and straightforward way.

Of course there are exceptions, and of course these are particularly interesting from a philosophical and sociological point of view, but to ignore the ordinary consensual experiments give a highly distorted picture of the role played by experiment in science. Moreover, the very fact that at least some experimental results are stable and constant over time and space seems difficult to explain from a constructivist perspective. The value of the gravitational constant has not changed much since the days of Cavendish, it has just become more precise; and the results do not depend the sligthest on the experimentalists' religion, nationality or cultural settings. Such permanence cannot easily be explained by the constructivist.

THE EDUCATIONAL CHALLENGE: TEACHING N-RAYS PHYSICS?

Does the constructivist view of science have any implications for the teaching of physics (and other sciences)? This question has been treated in detail by Peter Slezak in a sharp attack on contemporary sociology of scientific knowledge. As he points out, constructivism does indeed imply a serious educational challenge insofar that any insight in nature gained by experiment and critical thought is illusory according to constructivist doctrines: 'If... beliefs are not a matter of reasons, evidence and other rational considerations, then teaching cannot involve conveying ideas through understanding. ... the assumption that teaching involves the inculcation of ideas and understanding based on considerations of evidence and argument must be an illusion' (Slezak 1994, p. 267). Sociologists of knowledge have not shown much interest in science education, nor have they formulated an alternative to the present system of education. But there are enough of examples in the literature to indicate what a constructivist science education might look like.

To Collins and Pinch (1993), the value of science education is reduced to helping the students 'to do a lot of things such as repair the car, wire a plug, build a model aeroplane, use a personal computer to some effect, know where in the oven to put a soufflé, lower one's energy bills,' and so on (p. 150). Understanding, conceptual knowledge, and intellectual pleasure play no role. In an earlier work, Collins and Shapin (1983) consider it a key aim of physics education to teach the students that 'all experimental findings may be criticized, and no experimental finding need to be taken as a crucial confirmation or disconfirmation of a theory it is said to test' (p. 283). This pet doctrine of social constructivism implies that the experimental method should only be used in the classroom 'so long as it is the social organization of consensus that is the focus of attention rather than an abstract idea of the proper outcome of experiments' (p. 290). They suggest a deep-rooted analogy between the school laboratory and the laboratories of real science:

> Now and again, we suggest, students' attention might be drawn to the way the class, under the direction of the teacher, reduces an initially disordered set of quasi-findings into experimental support for a favoured (the correct!) hypothesis. The parallels between the classroom lesson and the life histories of scientific controversies as they move toward closure, are striking. (Ibid.)

Likewise, Collins and Pinch state confidently that 'every classroom in which children are conducting the same experiment in unison is a microcosm of frontier science' (p. 150). However, the parallel is more misleading than enlightening and tells very little about the nature of science.

For the sake of illustration, consider a typical classroom experiment, such as the determination of a pendulum's period T. The students are asked to find out how T depends on the length of the pendulum and we may suppose that they have not yet learned the formula derived from mechanics, that $T = 2\pi\sqrt{l/g}$ for small amplitudes. There will undoubtedly be discordant results and some confusion as to what to conclude. A few students may find that $T \propto l$, some that $T \propto l^{2/3}$, and (hopefully) most that $T \propto \sqrt{l}$. How should the teacher respond to this lack of consensus? Normally, the teacher will tell some of the students that their results must be wrong, check their methods and calculations, and ask them to redo the experiment. The result of guided criticism and improved experiments will quickly be that all the students agree that $T \propto \sqrt{l}$. When this has been established, the teacher may derive the corresponding formula from mechanics and discuss its limitations. Conclusion: It is a fact of nature that the period of a pendulum varies approximately as the square root of its length. This is not, however, an acceptable conclusion for the constructivist, according to whom truth and error are social and negotiable conventions. The constructivist teacher seems obliged to evaluate the $T \propto l$ and $T \propto \sqrt{l}$ results as equally valid and cannot use either theory or nature to discriminate between them. Within his framework, there are no errors in the conventional sense and no experiments that are intrinsically bad or good. 'How something comes to be seen as true, unambiguous and repeatable, is a social process,' he claims (Collins and Shapin 1983, p. 285) and so he cannot argue that $T \propto l$ is a wrong result because it disagrees with nature. Presumably, under the appropriate social conditions $T \propto l$ would be acceptable as a true and repeatable relationship.

It would be catastrophical to follow the constructivists in their relativistic conception of truth and error, good and bad science. It is an essential part of all education to learn from errors. In school physics we have the advantage that there are methods to identify errors and then to learn how to avoid them. There are good experiments and bad experiments, correct calculations and wrong calculations. In frontier science, anomalous results may be signs of a theory breakdown or a new discovery, but not in school physics. The student who concludes that $T \propto l$ or that the boiling point of water is 54°C has not found an anomalous result; he has made a plain error, which should be corrected. In short, if one drops the notion of error in an objective sense, how can physics be taught at all?

Another aspect concerns ethics and the norms of science. Constructivism implies that fraud, if only you can get away with it, is not necessarily bad or unethical. After all, so the constructivist explains, "fraud' is an outcome of discursive processes... [and] is to be seen as an attributed category, something made in a particular context which may become unmade later' (Pinch 1993, p. 368). According to this view, there is really nothing wrong with the student who fabricates data in his laboratory report or otherwise acts dishonestly.

Some constructivist sociologists are not satisfied with introducing social explanations as an alternative to natural explanations. A grotesque example of rampant relativism can be found in a constructivist account of the famous N-ray affair, an episode traditionally classified as pathological science. In 1903 the French physicist René Blondlot claimed to have detected so-called N-rays and his claim was supported by many contemporary physicists who produced confirming evidence for the feeble – and nonexisting – rays. As an example of how a spurious phenomenon acquired temporary respectability, the N-ray affair is instructive from the perspective of theory of science and is also relevant to science teaching. But according to the sociologist Malcolm Ashmore (1993), it is either wrong or meaningless to dismiss the rays as spurious. The remarkable thing is rather the consensus on the physical unreality of N-rays. Ashmore indicates that there can be no essential difference between the reality claims of X-rays, radioactivity, and N-rays and wants to 'inject a little healthy scepticism' in the standard accounts of N-rays. It is unclear if he really believes N-rays to exist, or once having existed, but apparently he and some other constructivists (Pinch 1993) find it all wrong to distinguish between accepted science, pseudo-science and discredited science.

Notorious cases such as N-rays, cold fusion, and polywater should therefore be reassessed and re-enter the world of science, including science teaching, on par with successful cases of science. 'I am looking for justice!' Ashmore proclaims in the name of relativism and the symmetrical framework of the Edinburgh school (Ashmore 1993, p. 70). It is difficult not to agree with the conclusion of Slezak (1994), namely, that Ashmore, Latour, Pinch and some other social constructivists advocate unorthodox and discredited science and find it 'unfair' that such work should not enter the science curriculum alongside respectable science. 'It seems that 'symmetry' has come to mean something like affirmative action for disreputable theories. Presumably, it follows that Blondlot should be given equal time with Röntgen in physics lessons' (Slezak 1994, p. 275).

All this must surely seem bizarre from the point of view of the physics teacher. Why on earth should he pretend that N-rays have the same 'right to existence' as X-rays when it is a plain fact that N-rays were a mistake, and X-rays were not. Any physics teacher worth his salt could influence his students (supposed to be sufficiently ignorant about the history of physics, which is a reasonable assumption) to believe in the reality of N-rays. I have tried the parallel case with phlogiston and had no trouble in persuading my students to believe in this imaginary substance. Such exercises can be pedagogically fruitful, but of course they should not be taken for more than that. New students of science can easily be fooled. And so what? Even experienced scientists can make mistakes, which should be no surprise. The point to emphasize is that errors in professional science can be corrected, and are in fact corrected. As a result of the critical methods employed in science we know for sure that whereas X-rays exist, N-rays do not. That ought to be reason enough to exclude N-rays from the curriculum; or, if including the subject, then treat it as a case-study of scientific error.

CONCLUSION: THE BABY AND THE BATHWATER

The positivistically coloured conception of science that is still, if only implicitly, the core of most science education is clearly unsatisfactory. The average student of a physics class is taught that physics is a collection of facts about nature, that these can be revealed by following the experimental method, and that consensus about truth is the hallmark of scientific work. The laws of physics are able to predict the outcome of experiments and are, at the same time, technologically useful and a key to unlocking nature's deepest secrets. How the vast body of physical knowledge has come into being is not part of the ordinary physics curriculum, which is also largely silent about the social, philosophical and political aspects of physics. There is surely good reason to broaden the scope of physics education and to present to the students a more realistic picture of how scientists work. Among these are the frequently occurring scientific and technological controversies where experts disagree. According to the received view, disagreement among scientific experts ought not to happen and must be accounted for in terms of individual incompetence or external, hence non-scientific, influence. As pointed out by Collins and Shapin (1983, p. 288), within the positivistic version of science controversies among scientists may therefore generate profound disillusionment about science and thus contribute to anti-science sentiments.

Social studies of science has led to much valuable knowledge and in general to a revised and more realistic picture of science. Elements of this knowledge will be relevant for the education of science and ought to enter the curriculum alongside Newton's laws and the properties of elementary particles. However, it is important to distinguish betwen social or cultural studies of science and the radical versions of constructivism and relativism. As argued by Slezak (1994), modern sociology of scientific knowledge must be wholly rejected and has nothing to offer the science teacher. It is philosophically unsound, has weak empirical support, and is subversive not only to good science education but to honesty and critical thought in general. If the received view of science may cause anti-science sentiments, constructivism is a frontal attack on the entire edifice of science and as such far more damaging.

Fortunately the physics teacher does not have to choose between the two extremes. Rejecting constructivism does not mean a rejection of the socio-cultural perspective of science or of specific findings of sociologists of science. Important as it is to avoid the strange blend of relativism and fundamentalism that characterizes modern constructivism, it is equally important not to throw out the baby with the bathwater.

REFERENCES

Ashmore, M.: 1993, 'The Theatre of the Blind: Starring a Promethean Prankster, a Phoney Phenomenon, a Prism, a Pocket, and a Piece of Wood', *Social Studies of Science* **23**, 67–106.
Brannigan, A.: 1981, *The Social Basis of Scientific Discoveries*. Cambridge: Cambridge University Press.
Cole, S.: 1992, *Making Science: Between Nature and Society*. Cambridge, Mass.: Harvard University Press.
Collins, H.: 1981, 'Stages in the Empirical Program of Relativism', *Social Studies of Science* **11**, 3–10.
Collins, H. & Pinch, T.: 1993, *The Golem: What Everyone Should Know About Science*. Cambridge: Cambridge University Press.
Collins, H. & Shapin, S.: 1983, 'The Historical Role of the Experiment', in F. Bevilacqua and P.J. Kennedy (eds.), *Using History of Physics in Innovatory Physics Education*. Pavia: The International Commission on Physics Education, pp. 282–292.
Dessaur, C.I. et al. (eds.): 1975, *Science Between Culture and Counter-Culture*. Nijmegen: Dekker and van de Vegt.
Franklin, A.: 1990, *Experiment, Right or Wrong*. Cambridge: Cambridge University Press.
Galison, P.: 1987, *How Experiments End*. Chicago: University of Chicago Press.
Gross, P.R. & Levitt, N.: 1994, *Higher Superstition: The Academic Left and its Quarrels with Science*. Baltimore: Johns Hopkins University Press.
Holton, G.: 1993, *Science and Anti-Science*. Cambridge, Mass.: Harvard University Press.
Kuhn, T.: 1992, 'The Trouble with the Historical Philosophy of Science', Robert and Maurine Rotschild Distinguished Lecture. Harvard University.
Labinger, J. A.: 1995, 'Science as Culture: A View from the Petri Dish', *Social Studies of Science* **25**, 285–306.
Laudan, L.: 1990, *Science and Relativism*. Chicago: Chicago University Press.
Murphy, R.: 1994, 'The Sociological Construction of Science Without Nature', *Sociology* **28**, 957–974.
Nowotny, H. & Rose, H.: 1979, *Counter-Movements in the Sciences: The Sociology of the Alternatives to Big Science*, Dordrecht: Reidel.

Pickering, A.: 1984, *Constructing Quarks: A Sociological History of Particle Physics*, Edinburgh: Edinburgh University Press.
Pinch, T.: 1993, 'Generations of SSK', *Social Studies of Science* 23, 363–373.
Sismondo, S.: 1993, 'Some Social Constructions', *Social Studies of Science* 23, 515–553.
Slezak, P.: 1994, 'Sociology of Scientific Knowledge and Scientific Education', *Science & Education* 3, 265–294, 329–355.
Stephenson, R.: 1982, 'The Skies of Babylon', *New Scientist* (August 19), 478–481.
Toulmin, S.: 1973, 'The Historical Background to the Anti-Science Movement', in *Civilization and Science*, CIBA Foundation: Amsterdam, pp. 23–32.
Trout, J.D.: 1994, 'A Realistic Look Backward', *Studies in the History and Philosophy of Science* 25, 37–64.
Weinberg, S.: 1992, *Dreams of a Final Theory: The Search for the Fundamental Laws of Nature*, New York: Pantheon.
Wolpert, L.: 1992, *The Unnatural Nature of Science*, London: Faber and Faber.
Woolgar, S.: 1983, 'Irony in the Social Study of Science', in K. Knorr-Cetina and M. Mulkay (eds.), *Science Observed*, London: Sage, pp. 239–266.

Coming to Grips with Radical Social Constructivisms

D. C. PHILLIPS

School of Education and Department of Philosophy, Stanford University, CA 94305, USA

ABSTRACT: This essay distinguishes two broad groups – psychological constructivists and social constructivists – but focusses upon the second of these, although it is stressed that there is great 'within group' variation. More than half of the paper is devoted to general 'clearing of the ground', during which the reasons for the growing acrimony in the debates between social constructivists and their opponents are assessed, an important consequence of these debates for education is discussed, and an examination is carried out of the radical social constructivist tendency to make strong and exciting but untenable claims which are then backed away from (a tendency which is documented by a close reading of the early pages in Bloor's classic book). The last portion of the essay focuses upon social constructivist accounts of the causes of belief in science – the more radical of which denegrate the role of warranting reasons, and which give an exalted place to quasi-anthropological or sociological studies of scientific communities.

If an unsuspecting researcher was to carry out a computer search for articles across the fields of education, sociology, psychology, and epistemology that used the term 'constructivism', the results would be overwhelming. And also quite bemusing – for it would turn out that the term is used almost without rhyme or reason. The situation has become so confusing that to be told that a particular individual is a 'constructivist' is to acquire no useful information whatsoever.

In an earlier essay (Phillips 1995) I tried to bring some order to this troubling situation by delineating two very broad constructivist 'camps' each of which had substantial 'within group' differences; I labelled these (1) 'psychological constructivists' (Piaget, von Glasersfeld, and even Vygotsky would be included here, and the empiricist philosopher John Locke would be a borderline figure), and (2) 'social constructivists'. Members of the first group have, as the focus of their interest, the 'constructions' or 'cognitive and memory structures' or 'understandings' in the mind of the individual learner or knower, and how these are built up. Members of the second group are concerned with the public bodies of knowledge – the disciplines – and how these are constructed over time; they downplay the role played by 'external reality' in shaping our beliefs, and instead they stress (to varying degrees) the role played by social processes within knowledge-producing communities. Most of my efforts on that previous occasion went into clarifying certain features of the first group, the psychological constructivists, so here I propose to focus upon the second. But before we can grapple with any of the interesting theses put forward by one or other of the social constructivists (actually I will only have room to deal with one such thesis), there is a great deal of preliminary work to be done.

CLEARING THE GROUND

1. It is notable that the quite remarkable spread of social constructivism over the past two decades has not taken place quietly – the debates between constructivists and their critics have been pursued with vigor and with growing heat. Exchanges were lively, but relatively polite, in the early 1980's. For example, there was a notable lengthy debate in the pages of the journal *Philosophy of the Social Sciences* between the philosopher of science Larry Laudan whose contribution had the provocative title 'The Pseudo-Science of Science?', and David Bloor – a central member of the 'strong programme in the sociology of knowledge' centered in Edinburgh – whose lively response had the misleadingly meek title 'The Strengths of the Strong Programme' (see Laudan 1981; Bloor 1981.) However, the follow-up debates spawned by these papers quickly became, in the words of one commentator, 'acrimonious'. (Fuller 1993, p. 12; many of the relevant papers were collected in Brown 1984.)

By a little more than a decade later the emotional temperature had risen even further. A variety of interesting events occurred in the first half of the 'nineties, ranging from vituperative exchanges at conferences (Times Higher Education Supplement 1994, p. 44; Keller 1995, p. 14), to publication of fiery books and reviews (Gross and Levitt 1994; Cole 1995; Lovie 1995; Nussbaum 1994; New York Review of Books 1995). Even Thomas S. Kuhn, whom many regard as one of the great ancestral figures of contemporary social constructivism, entered the fray and stated in a public lecture at Harvard that the claims of the Edinburgh School are 'absurd: an example of deconstruction gone mad'. (Kuhn 1992, p. 9)

The discussion above must not be taken as suggesting that there are no books and essays, on either side of the constructivist divide, where name-calling and epithets are kept to a minimum; but as one reads in this general field it is difficult to avoid reaching the conclusion that strong feelings are bubbling just beneath the surface. And the reason for this is not difficult to discern: *There is a lot at stake*. For it can be argued that if the more radical of the sociologists of scientific knowledge (not to mention a variety of postmodernists and some feminist epistemologists) are right, then the validity of the traditional philosophic/epistemological enterprise is effectively undermined, and so indeed is the pursuit of science itself. In the most radical scenario, epistemology has all the validity and relevance to the modern world as medieval alchemy; in more moderate scenarios epistemology at least has to be reconceptualized as part of – but not a special or authoritative part of – what Rorty called the 'conversation of mankind' (Rorty 1979). The picture is even bleaker for science, which according to many of the social constructivist and postmodernist accounts becomes little more than an exercise in ideology building (in the pejorative sense of this expression). As Cleo Cherryholmes summarized the case, scientific research is a species of practice, and 'human interests, myths, ideologies, values, and commitments shape what researcher-theorists claim to know...' (Cherryholmes 1988, p. 111); it is noteworthy that he does not list nature itself as playing any role in shaping or constraining what scientists believe about it! Or as Latour and Woolgar put it, in their famous constructivist study *Laboratory Life*,

> there is little to be gained by maintaining the distinction between the "politics" of science and its "truth"...the same "political" qualities are necessary both to make a point and to out-manoeuvre a competitor. (Latour and Woolgar 1986, p. 237)

Evelyn Fox Keller offers another diagnosis of why the exchange has been so heated, particularly on the part of those who favor a traditional ('modernist' or 'Enlightenment') view of science: In these times when research is incredibly expensive,

> the basic fact is that science – especially big science – is in a position of utter and absolute dependency. Without the continuing support of the nonscientific public, the institution of science as we have come to know it, and as working scientists have come to take for granted, will end. It is hardly surprising that scientists would want the public to perceive their ventures in the best possible light.... From the anxious perspective of scientists alarmed by recent trends in funding, science studies comes to look like the enemy within. (Keller 1995, p.15)

I have pointed out, in my earlier paper, another factor that might explain the intensity of the debates: The constructivists generally have strongly held socio-political motives (for example, the need to make scientific research communities more inclusionary) that undergird their work and seem to be related to – if not to drive – their analyses of the production of knowledge. (Phillips 1995).

2. I mentioned earlier the speciation and formation of sects that has taken place within the broad social constructivist camp; the 'within group' variation here should not be underrated. Thus the social constructivists form a truly Wittgensteinian family with few if any common threads linking the intellectually widely-dispersed members. Marx presumably must be credited with being one of the family's distant ancestors, for he held the view that the ruling ideas/beliefs in a society are those of the ruling class, and serve its (economic) interests. That is, beliefs or ideas are tools or weapons in the warfare between the classes in society. The constructivist picture here is somewhat muddied by the fact that Marx traced the causal chain back to the material/economic base of society, and he also allowed for the fact that acting upon ideas can reflexively produce changes in that material and economic base.

But if Marx is the base drummer in the social constructivist orchestra, there are many other instrumentalists: Emile Durkheim, Mary Douglas, Thomas Kuhn, Richard Rorty, Michael Mulkay, Barry Barnes, David Bloor, Harry Collins, Stanley Fish, Bruno Latour, Michel Foucault, the three Steves – Fuller, Woolgar, and Shapin, and a diverse group of feminist thinkers including Sandra Harding, Helen Longino, Lynn Hankinson Nelson, and Evelyn Fox Keller. These folk surely constitute a strange ensemble, for they are not playing the same tune, nor are they following the beat of the same conductor. Some (such as Kuhn) even have tried to resign from the orchestra (see Kuhn 1992); and a few marginal figures can be discerned (such as Goldman and Kornblith) who keep one foot firmly planted on traditional epistemological ground while they gingerly test moderate constructivist waters with the toes on the other foot.

For those who like more order than is apparent in this list, a number of rough groupings of social constructivists can be identified, ranging from those whom Latour (in an interesting paper 1992, p. 276) called 'radical', through 'progressive', and all the way to a 'marsh' of 'wishy-washy' scholars; Latour used as his categorizing principle the degree to which it was insisted that 'external nature' played an important role in shaping the 'knowledge' that was produced. In more conventional terms, there are (roughly) the members of the strong programme, the scholars in the field known as social studies of science (sometimes also called 'science and technology studies'), the postmodernists and an overlapping group of feminist epistemologists, and finally the group of moderate and even 'conservative' constructivist philosophers (see Kornblith 1994) who work in the domain of 'naturalized epistemology'. But even within these groupings, there is a great deal of individual variation. In a recent book Sergio Sismondo separates out six distinct uses of the 'construction metaphor' (although he recognizes that there are more), and he writes that

> Even though social constructivism and constructivism are popular as approaches to science, it is sometimes unclear exactly what is claimed is constructed and how. .... "social construction" and "construction" do not generally mean the same thing from one author to another and, even within the same work the terms are meant to draw our attention to several quite different types of phenomena. (Sismondo 1996, p. 49)

As most of these groups and individuals – however we characterize them – focus their discussions on the social construction of *scientific* knowledge, I shall follow their example in the present paper.

3. There is an issue which seems to me to be so clear-cut that I am reluctant to raise it; but, as I have occasionally been questioned about this, it is as well to state my position explicitly. The question can be asked: 'Of what relevance to education are the debates over social constructivism?' My reply is set out in the following paragraphs.

Teachers of the natural sciences, math, psychology, history, economics, and so forth – both in high schools and in universities – often want their students to acquire more than just an understanding of the findings in these fields; they also want students to appreciate the nature of these fields, in the sense that they want students to be acquainted with *how knowledge is built up or acquired* in these fields, how it is established or tested or how it attains its status as knowledge; and certainly textbooks often give some account (usually a truncated and half-hearted one) of these matters. To make the point in the terminology made famous in the 1960's by Paul Hirst and Joseph Schwab, educators often expect students to understand not only the substance or 'substantive structure', but also the tests against experience and the techniques – together, the 'syntactical structure' – of the subjects they are studying. (For a discussion of Hirst and Schwab, see Phillips 1987, ch. 11)

Now, if the views of the social constructivists are correct, we need to revise thoroughly the account we give to students of the syntax of the disciplines they are studying. Physics – to take an example where the issue emerges rather starkly – should no longer be depicted as being the quest for a true account of the objectively determinable properties and forces in the pre-existing natural realm which has reality independent of what humans happen to believe about it at any particular time, this quest being loosely guided by a rationally defensible 'scientific method'.

The precise story to be told in place of this traditional account will differ according to which social constructivist you listen to, but might well be along the following lines which, among other things, problematizes the notion of an independently-existing 'nature' (this crude 'thumbnail sketch' will be rounded out more carefully in the subsequent discussion, where some at least of the philosophical issues will be laid out more clearly): The natural realm is not pre-existing but rather is constituted by our inquiries, and rather than being driven by a rational 'scientific method' our inquiries take the form that they do because of various social factors and processes; and it follows from this that 'the [pre-existing] natural world has a small or non-existent role in the construction of scientific knowledge' (Collins 1981, p. 3). As Woolgar puts it, there is 'an inversion of the relationship between objects in the world and their representation. It was suggested [in his earlier discussion] that representational practices *constituted* the objects of the world, rather than being a *reflection* of (arising from) them'. (Woolgar 1993, p. 67, emphasis added)

Some at least of the social constructivists would not hesitate to draw some philosophical 'morals' from this, which in all likelihood also would become part of the 'meta-narrative' that they would substitute for the traditional account of the nature of disciplines such as physics. Steve Fuller writes (concerning the beliefs of another noted radical social constructivist, David Bloor): 'Echoing the later Wittgenstein, Bloor holds that science can be explained as a form of life without importing conceptions of truth, rationality, and reality that require special philosophical grounding'. Indeed, Bloor and his allies 'argue that the appeal to philosophical concepts has largely had political, not scientific, import – to consolidate allies and to exclude rivals' (Fuller 1993, p. 12). Rorty is even blunter: We should 'see knowledge as a matter of conversation and of social practice, rather than as an attempt to mirror nature . . . '. (Rorty 1979, p. 171)

In short, if one or other of the more radical social constructivist accounts was to be accepted, physics would no longer be depicted in our textbooks and classrooms as an enterprise which seeks true accounts of external reality; physics instead would be depicted as a political enterprise or as a realm of Rortyean conversation – a conversation that, explicitly, is *not* shaped to any significant degree by external nature, and by warranting reasons. Later I shall discuss the flawed view of how humans act and settle upon beliefs that undergirds the social constructivist position here.

4. In his book *The Rationality of Science* (1981), the British philosopher W. H. Newton-Smith has a detailed discussion of Thomas S. Kuhn's work on scientific revolutions. He noted that Kuhn starts by making claims that are strong, exciting, but false; however, when under pressure, Kuhn backs off and softens them so that they become more credible, but weak and boring. I claim that the same phenomenon can be noticed at crucial places in the work of many social constructivists. (In the terminology used by the social constructivist Bruno Latour 1992, there is a tendency for those who initially hold a radical social constructivist position to move towards progressivist or even 'wishy-washy' positions. (Forman 1995, argues a similar point in detail.)

In a volume with the revealing title *Rethinking Objectivity* (Megill 1994), one of Bloor's colleagues at the center of the 'strong programme', Barry Barnes, writes that sociologists of knowledge

> should now beware of overshooting the mark. Several of the most significant current difficulties and weaknesses in the field are over-reactions to the individualism, rationalism, and realism typical of so much epistemology. (Barnes 1994, p. 27)

At first blush it seems as if Barnes has significantly softened his earlier relativistic social constructivism: The weaknesses in the position are due to over-reacting to realism and so forth. On the assumption that one is either a realist (of some form or another) or one isn't, the only way an anti-realist can stop 'over-reacting' is by *becoming* a realist! But, in this case, social constructivism is on the road to becoming a boring position, for the door has been opened to allowing that reality or 'nature' plays *some* role in shaping the beliefs that are held about it. This latter view is one that – in my estimation – is both true and boring. (It is boring in the following sense: If members of the strong programme had said this in the first place, their works would have generated little if any controversy, the arguments would not have taken place, and the heat would not have been generated. Furthermore, the present paper would hardly have needed to be written!)

But there is a further point to be made here: Barnes was attempting to have his cake, and eat it too. For, immediately preceding the passage just quoted, in which he called for the abandonment of '*over-reaction*', Barnes suggested that 'sociology of knowledge was correct in its *uncompromising rejection* of [pre-existing] epistemology' (Barnes 1994, p. 27, emphasis added). So we have the following position being canvassed, one in which the exciting element is first reaffirmed, then denied and modified in a boring direction that – if taken seriously – completely undermines what has just been reaffirmed: Traditional 'individualistic' and 'realist' epistemology should be uncompromisingly rejected, and yet it is a mistake for social constructivism to 'overshoot' the mark and 'over-react' to this epistemology about which, nevertheless, we are urged to be uncompromising!

If I may be forgiven for making a sociological point in the midst of a philosophically-oriented discussion, one very important function is served by Barnes's shilly-shallying here – he has given himself a route by which to escape from serious criticism. For if one were to attack his rejection of realism and so forth, he can say this is beside the point as he himself has pointed out that such rejection is an over-reaction; and yet if one were to criticize his new flirtation with realism, he can also argue that this is beside the point as he has reaffirmed that it is correct to be 'uncompromising' about it. In effect, Barnes is illustrating that a position that is internally inconsistent has some significant advantages in social settings that do not put much emphasis on the importance of logical consistency.

In an essay in 1981 another strong sociologist of knowledge, Harry Collins, whether wittingly or not, adopted a variant of the same strategy – he made a very strong claim in one paragraph, and on the very next page softened it to such an extent that it became innocuous. Collins was writing an introduction to a collection of work in a special issue of a journal (which included, *inter alia*, one of his own papers); all of this work had been carried out in what he labelled as 'the empirical programme of relativism', and he described it in these terms:

> One school, however, inspired in particular by Wittgenstein and more lately by the phenomenologists and ethnomethodologists, embraces an explicit relativism in which *the natural world has a small or non-existent role* in the construction of scientific knowledge. One set of such analyses is gathered in this issue of *Social Studies of Science*. (Collins 1981, p. 3, emphasis added)

However, Collins went on to state that 'All the papers confirm the potential local interpretative flexibility of science which prevents experimentation, *by itself*, from being decisive' (Collins 1981, p. 4). The words I have emphasized here mark a significant shift; to say that experimental results, by themselves, do not decisively shape the nature of scientific belief (which is in my view a true but unexciting moderate position), is a far cry from the exciting (but ultimately incredible) claim that *no* experimentation, *no* input from nature, plays a role in shaping scientific belief. (I should note that the notion of 'nature' is held by some constructivists to be itself socially constructed; this is true, in a sense, but it does not seem to allow constructivists to extricate themselves from the problems here – for one thing, it still begs the question of what role external reality plays in influencing what we contruct about it.)

The issue that arises here is a critical one for the remainder of my discussion: Which theses – the exciting but incredible ones, or the moderate, wishy-washy and boring – should be the focus of attention when evaluating the positions held by various social constructivists? My strategy in the final sections of this paper will be to opt for excitement, on the ground that it is the exciting versions of social constructivism (rather than the dull ones) that have been of widespread influence over the past two decades.

Before turning to a more detailed examination of one of these exciting ideas held by the more radical of the social constructivists, however, there is one last preliminary issue to be discussed. This will necessitate returning to the issue of having one's cake and eating it too.

5. In his book *Science: The Very Idea*, the social constructivist Steve Woolgar noted that, to most minds, there is an 'apparent absurdity' to the radical social constructivist position (Woolgar 1993, p. 67); he put this down to the strong hold that the traditional account of science has upon most of us. The point is, however, that the radical social constructivist position *is* absurd, and does not suffer merely from *apparent* absurdity.

But here, of course, another interesting issue arises: How is it that intelligent people, like Bloor and Woolgar and the rest, can accept a general position that is so absurdly counter-intuitive? My own hypothesis about this is as follows: Bloor (to stick with him as a major foundational figure) over-states his case in the early pages of his book, and then later modifies his claims in a moderate direction, without recognizing that he has thereby shot himself in the foot! When attacked, as for example in the exchange with Laudan in 1981, his defence makes use of his moderate statements; but when speaking at a very general level he reverts to using the earlier absurdities. The point is, he cannot let these defective claims go, for what would a 'strong programme' be without its strong claims? In short, Bloor attempts to keep his radical cake, but when pushed he starts to eat it in wishy-washy company.

(a) Consider first the radical claims he makes in the opening pages of the first chapter of his famous book. I have strung together here a few short excerpts, together with my comments; you will need to read the chapter for yourself to see that I have not misrepresented him by using the method of selective quotation. (All the passages are from Bloor 1976; I shall just cite page numbers. In my comments I shall stick with the example of physics).

Quotation: 'Can the sociology of knowledge investigate and explain the very content and nature of scientific knowledge?' (p. 1)

Comment: This is a rhetorical question, for Bloor goes on to say that those sociologists who answer the question in the negative are guilty of 'betrayal of their disciplinary standpoint'. Why does Bloor's position strike many of us as absurd? Because we think that the content of the science of physics is explainable (at least to a significant degree) by physical considerations – by the theories, experimental and observational evidence, mathematical derivations and so forth that give physicists grounds for believing that acceptance of the current content of their field is warranted. Note that this is not to say that other factors are not part of the story; but to suggest that physics-type reasons (what some writers refer to as the 'internal reasons') are *no* part of the story (Bloor, after all, does not mention them here) is to suggest that we could replace physicists by sociologists – the physics of quantum mechanics is replaceable by the sociology of quantum physicists.

Quotation: 'The cause of the hesitation to bring science within the scope of thorough-going sociological scrutiny is lack of nerve and will'. (p. 2)

Comment: On the contrary, it is the result of good sense: for what Bloor means by 'thorough-going scrutiny' is outrageous. There is, of course, an important place for the sociological study of physics and physicists, but sensible sociologists should resist the temptation to explain the *content* of physics without looking at – among other things – the scientific/physical evidence adduced by physicists.

Quotation: 'Similarly, the sociologist seeks theories which explain the beliefs which are in fact found [in science], regardless of how the investigator evaluates them'. (p. 3)

Comment: This merely repeats the previous mistake. In physics (as indeed in most rational human endeavors), an important part of the story accounting for why beliefs are accepted is the fact that the people involved hold that there is warranting evidence in favor of those beliefs. (Physicists, as well as other mere mortals, may of course be mistaken in the evaluation of the warrants for their beliefs; but to pursue this issue, the warrants themselves need to be examined.)

Quotation: 'the sociology of scientific knowledge should adhere to the following four tenets.... These are: 1. It would be causal, that is, concerned with the conditions which bring about belief or states of knowledge. Naturally there will be other types of causes apart from social ones which will co-operate in bringing about belief'. (pp. 4–5)

Comment: At first sight this looks hopeful, for Bloor acknowledges 'other types of causes'. However, nowhere in these strongly-worded pages does he mention that the 'internal' or scientific reasons or evidence must be included among these 'other causes'. It is worth noting that postmodernists fill out Bloor's list by adding political, economic, ideological, and perhaps psychological causes to his 'sociological' ones; also, as is well-known, Lyotard is 'incredulous' about the justificatory metanarratives – in terms of truth, evidence, and so forth – that are told in an attempt to warrant belief in the universal claims of science (see Lyotard 1984). Bloor is not alone.

Bloor continues on with his list of 'tenets'; the second one reinforces the view that he is not concerned with the reasons or evidence that might be adduced within physics in support of the beliefs of physicists – for the truth or falsity of beliefs is not to be taken into consideration.

Quotation: '2. It [sociology of science] would be impartial with respect to truth or falsity, rationality or irrationality, success or failure'. (p. 5)

Comment: Case made – Bloor's position, as stated here in the opening pages of his first chapter, does have more than an *appearance* of absurdity. His position is absurdly 'strong'.

(b) We now need to move a few pages further into Bloor's book, where – despite his strong earlier claims – he becomes moderate and wishy-washy. For, suddenly, he allows that *experience* plays a role in science! In one powerful passage he writes:

> No consistent sociology could ever present knowledge as a fantasy unconnected with men's experiences of the material world around him [sic]. Men cannot live in a dream world. (p. 29)

Worse still for his earlier position, within a few lines he acknowledges 'the reliability of perception and the ability to detect, retain, and act upon perceived regularities and discriminations'. (p. 29) But why, then, did he not allow that reliable perceptions and discriminations are part of the evidentary material that convinces physicists to accept the things (the content of physics) that they do?

A hint about the answer to this question can be found in the closing pages of Bloor's first chapter. Here he makes a not unreasonable point (one amply endorsed in mainstream philosophy of science):

> But theories and theoretical knowledge are not things which are given in our experience. They are what give meaning to experience by offering a story about what underlies, connects, accounts for it. This does not mean that theory is unresponsive to experience . . . . [But] *Another agency apart from the physical world is required to guide and support this component of knowledge.* (pp. 12–13, emphasis added.)

On a straightforward reading, to admit that 'another agency apart from the physical world is required' is to allow that 'the physical world' is *part* of the story that must be told to account for the beliefs of physicists. In the rest of this passage, Bloor is quite right: theory is *underdetermined* by experience/nature (see Phillips 1987, Part A). This means that it is possible to erect alternative theories to account for our experience, our scientific data. But Bloor has made a monumental error in supposing, then, that because alternative theories can be produced, it somehow follows that experience/nature can be *left out of the account* of why physicists believe the things they do.

A simple thought experiment can illustrate this: Suppose that the experience of physicists had been quite different (than it actually was) when they were experimenting with magnets, or with electric currents; or suppose that balls rolling down inclined planes did not (and do not) act in the manner which Galileo and countless subsequent physicists have observed. Clearly, the nature of the contents of the discipline of physics would now be quite different from what it is – for because the balls (or whatever) would have been behaving differently, the beliefs we arrived at would have been different. One of the causal factors helping to shape the contents of physics is the evidence or experience that physicists obtain when experimenting or observing, even though this evidence alone does not fully determine the nature of the theories that can be developed to explain it.

The moral is that external nature cannot be conceptualized in any way that we please – as Bloor seems to acknowledge. (This point is argued strongly by Israel Scheffler 1967.) If Bloor is serious about the points he is making in these later passages, he cannot consistently leave untouched the absurd remarks he made in describing the essence of the strong programme in the opening pages of his book. The strong programme has to become a wishy-washy one (to borrow Latour's expression) in order to be credible.

There is an even deeper fault underlying Bloor's work: He seems to have an unacceptable model of what it is to be human, of what factors cause humans to act and to adopt beliefs. For his strong words do not acknowledge that humans are rational animals. I will turn to this central point in the discussion below.

EXPLAINING BELIEF

There is a (wishy-washy) sense in which it is entirely non-controversial to claim that knowledge is socially constructed: A scientist will be working on a problem that is describable in a public language (even if this is a technical language), and it will have emerged *as* a problem in the context of a shared conceptual/theoretical framework. Furthermore, he or she will be using a set of communally-endorsed practices and tools – concepts, apparatus, research designs, data analysis techniques, forms of inference, and the like, that are familiar to (and which probably have been developed by) others in the same professional specialization. The evidence that is collected either will be of a type that is accepted as relevant to the issues at hand by specialist colleagues (who themselves constitute a socially-defined group), or else it will be evidence that the inquirer can argue is relevant and worthy of acceptance. The knowledge-claims that result from this work will be expounded in papers that are examined by journal referees, conference organizers, respondents at meetings, and so on, and attempts at replication will most likely occur. A claim that survives this scrutiny, and that is judged as important or that is widely used by others in the speciality as a basis for their own work, might eventually be written up in textbooks and become part of the curriculum in schools and universities. In these important senses, then, knowledge is a social product; an inquirer working *entirely* on his or her own, using no social resources *whatsoever*, is a ludicrous fiction.

Because knowledge production is a social phenomenon in the above sense, it follows that sociologists and anthropologists and psychologists are right to stress that it can be studied in the manner in which other social phenomena can be studied – namely, empirically, using the methods of the social sciences in an attempt (perhaps) to delineate at least some of the causal processes involved. To put it bluntly, the production of scientific knowledge can itself become the subject of scientific scrutiny – a statement that has the air of paradox although in fact there is nothing paradoxical about it at all. Insofar as they are saying this, Bloor, Woolgar and others are on firm ground. Indeed, Bloor expressed this so-called 'causal principle' in his early formulation of the 'strong programme', and in the following passage there is nothing at all that I would want to disagree with: The sociology of knowledge

> would be causal, that is, concerned with the conditions which would bring about belief or states of knowledge. Naturally there would be other types of causes apart from social ones which will cooperate in bringing about belief. (Bloor 1976, pp. 4–5)

There are three crucial issues, however, that arise concerning this social-scientific program: (a) What are these *'other types of causes'* that can be (or, should be?) invoked to explain the social construction of knowledge? (b) What *aspect* of the construction of knowledge can be explained in sociological terms? and (c) Are knowledge-claims that are *true* (or judged to be true) explained in different terms from those that are *false*?

(a) Types of causal factors. Bloor himself noted that the causes that could be invoked to explain the construction of knowledge, in addition to sociological ones, were psychological, political or historical; and many postmodernists would agree with him. We saw earlier that Cherryholmes gave a comparable list, and Latour and Woolgar were approaching close to Bloor when they argued that the study of scientific truth should not be distinguished from politics. Presumably the point here is that a scientist's tenacious pursuit of a particular problem (to take a simple example) might be explainable in sociological terms (the scientist might be displaying the behavior or interests typical of a member of a particular socio-economic class), or in political terms (the scientist might be under pressure from some powerful figure such as the head of the laboratory), or in economic terms (he or she stands to make a handsome profit), or in psychological ones (there might be Freudian undertones in the scientist's behavior).

And, of course, these are not mutually exclusive, for many phenomena are overdetermined or multiply caused. But it is notable that what is excluded by the radical social constructivists (or at least what is not explicitly mentioned) is the *content* of the science that is being wrestled with, and the status of the warrants or supporting arguments or evidence for the beliefs that are being taken seriously by those scientists; in the earlier discussion Bloor's shilly-shallying over this issue was discussed at some length (it will be recalled that different portions of his classic book take contradictory stands on the role of a scientist's experience).

The position held by Latour and Woolgar is also revealing. In the early pages of *Laboratory Life* they discussed what their approach should be to studying the construction of knowledge in the scientific laboratory in which Latour's data were collected, and they ended up *rejecting* an 'emic' orientation wherein they would have to try to understand the meaning (for the participants or subjects they were studying) of the terms and ideas that these individuals were using in their knowledge-producing activities. Latour and Woolgar took this decision on grounds that are difficult to comprehend and which have something of the flavor of 'gobbledy-gook': They say they were aware of the dangers of 'going native' (which apparently would have been an issue if they had adopted the emic approach and had attempted to understand their scientists, although Latour and Woolgar are far from clear on this vital point), and so they decided to treat the concepts of the participants 'as a social phenomenon' (Latour and Woolgar 1986, p. 39). The point is, that by whatever strange logic, Latour and Woolgar ended up studying how knowledge is produced, by adopting the perspective of outsiders – that is, *without* taking into account the *meaning* of the conceptual/theoretical material the scientists in the laboratory were discussing, questioning, and taking as their framework, and in terms of which these individuals were expressing the very knowledge they were constructing!

The enormity of the position adopted here by Latour and Woolgar cannot be underscored sufficiently; and clearly their stance is quite at variance with the one that is adopted by mainstream ethnographers who 'bend over backwards' to understanding the point-of-view of the groups they are studying. One consequence of the strange methodological decision of Latour and Woolgar is this: Having decided *not* to take into account the *content* of the discussions, arguments, and papers that occupied much of the laboratory scientists' time, and by narrowing their attention to only the observable socio-political interactions (and suchlike) in the laboratory, Latour and Woolgar were ensuring that they would 'discover' that knowledge construction is 'political' in nature and that issues of 'truth' are irrelevant and play no causal role. In the 'Postscript' to their book's second edition they even tried to pass off as an advantage the fact that when Latour started the fieldwork, all of which he carried out without Woolgar's assistance, his 'knowledge of science was non-existent; his mastery of English was very poor; and he was completely unaware of the existence of the social studies of science'. (Latour and Woolgar 1986, p. 273) Given that fieldworkers usually are expected to be aware of the ways in which their own backgrounds can bias their 'findings' – a point that social constructivists, postmodernists, and feminist epistemologists often wisely are sensitive to – this is an astounding admission. Peter Slezak had a similar reaction to the one I have just expressed:

> On the face of it, the author's own description of their project in *Laboratory Life* reads more like a parody than a serious inquiry ... the idea that the inability to understand one's human subjects is a positive methodological virtue is surely a bizarre conception even for anthropology. For Latour and Woolgar, however, it is intimately connected with their doctrine of 'inscriptions'. The meaninglessness of the 'traces, spots, points' and other recordings is a direct consequence of Latour's admitted scientific illiteracy. (Slezak 1994, Part 2, p. 336)

Slezak points out that Latour and Woolgar, among other failings, do not appreciate the difference between understanding the people they are studying, and believing them; their approach thus turns 'incomprehension into a methodology'. (Slezak 1994, Part 2, p. 336)

(b) What aspects of knowledge construction are being explained? It is a principle widely accepted in the social sciences that the appropriateness of the methods adopted by researchers should always be evaluated in the context of the questions that they are attempting to answer. In light of this principle, I am not suggesting that it is always necessary for investigators to understand what those they are studying are saying or believing. It all depends upon what it is that the researchers concerned are trying to elucidate. The 'methodology of incomprehension' adopted by Latour might be appropriate for *some* sorts of inquiries. An example of a research tradition that was fruitful, and where there was a conscious effort *not* to pay attention to the meanings and understandings of the people being studied, is the work done on the interaction patterns that occur during conversations; researchers here concentrated not on the content of the conversations they were studying, but rather focussed (for example) on the timing of remarks, the length of pauses, and the other non-verbal signals by which individuals in the conversations established their respective turns to speak.

However, Latour and Woolgar were not chiefly interested in matters such as this; together with others involved in the social studies of knowledge (including of course the members of the Edinburgh School) they were concerned to account for the production of the actual *content* of science and other domains of knowledge. Writing in a later book, Woolgar stated:

> It should be clear from these tenets [of the strong programme, of which he is supportive] that mathematical statements such as '2 + 2 = 4' are as much a legitimate target of sociological questioning as any other item of knowledge .... What kinds of historical conditions gave this expression currency and, in particular, what established (and now sustains) it as

a belief? This kind of question is posed without regard for the (actual) truth status of the statement.... (Woolgar 1993, p. 43)

But surely it was strange for him to *exclude* the 'truth status' or the theoretical arguments within mathematics from being elements giving '2 + 2 = 4' its 'currency'! (Certainly during the course of history views of mathematicians have changed about the way statements like '2 + 2 = 4' should be analyzed; but to discuss and assess these views, and their changes, the *content of mathematics* has to be explored – mathematics as a discipline cannot be replaced by sociology of knowledge!)

Woolgar revealingly concluded, a few pages later, that 'a central achievement of the sociology of scientific knowledge is its scepticism about the role of logic and reason especially in mathematics and science' (Woolgar 1993, p. 50) – an 'achievement' that depended entirely upon Woolgar and his fellow-travellers *legislating* or *assuming* at the outset that these factors played no causal role. Many years ago Bertrand Russell, looking at different research traditions to those that concern us here, noticed this same phenomenon:

One may say broadly that all the animals that have been carefully observed have behaved so as to confirm the philosophy in which the observer believed before his observations began. Nay, more, they have all displayed the national characteristics of the observer. Animals studied by Americans rush about frantically, with an increasing display of hustle and pep, and at last achieve the desired result by chance. Animals studied by Germans sit still and think.... To the plain man, such as the present writer, this situation is discouraging. (Russell 1948, pp. 32–33)

To which one can add: Scientists studied by sociologists from Edinburgh or France spend their days in laboratories manipulating meaningless symbols and making meaningless inscriptions.

(c) Should the knowledge claims that are regarded as true be explained differently from the ones that are false? It is clear that the radical social constructivists I have been paying most attention to in this paper do not wish to allow the distinction between 'true' and 'false' to have any purchase – for if they did allow that this distinction was causally relevant, they would be forced to consider, and assess, the grounds upon which judgments of truth and falsity were made. Latour, for one, would have been sorely at a disadvantage if his investigations had needed to move in this direction – his behavioristically oriented methodology was not able to deal with *reasons*. But it is not only Latour who dismissed this distinction; Bloor had included as the third of his four central tenets of the strong programme a principle of 'symmetry': work done within the programme 'would be symmetrical in its style of explanation. The same types of cause would explain, say, true and false beliefs'. (Bloor 1976, p. 5)

Some background here might make the position of the social constructivists somewhat more understandable. They were reacting against a view, advocated by a number of philosophers, to the effect that the acceptance of true propositions by scientists called for no (further) explanation, but what was needed was an account of why researchers accepted false ideas. Imre Lakatos argued particularly strongly for this position. (Lakatos 1978)

Lakatos drew a distinction between internal and external accounts of science; and he argued that since science is an intellectual, knowledge-producing endeavor, internal accounts must take precedence over external ones. The general idea seems to be this: Internal accounts focus upon the intellectual problems that scientists were struggling with, the evidence that was available, the theories that were held at the time, the standards of evaluation that were judged to be warranted and appropriate to the situation, and so forth. Now, if a scientist was to accept an idea that, by these internal standards and so forth, was judged to be true, the scientist would be behaving rationally (given the context) and no further explanation would be required (for to say that a person was acting rationally in a given situation is to say that the person was acting appropriately). If however, given the preponderance of the internal evidence and so forth, the scientist was to accept an idea that was false (judged by these standards), then his or her behavior was in need of further explanation – for now the scientist was not doing what seemed to be required of a rational person. On these sorts of occasions, Lakatos was arguing, explanations should be sought externally; perhaps the scientist had been swayed by political or religious factors, or was guilty of sexist or racist bias, or so forth.

Here, I think, the social constructivists were guilty of misinterpreting the position of Lakatos and others (whether this was fueled by incautious wording of the philosophers' theses is another issue). The constructivists took it that what was being claimed by the philosophers was that true belief needs no causal explanation, whereas their own position was that all belief should be subject to causal explanation. Hence Bloor formulated his 'principle of symmetry'; and Woolgar commented favorably on 'the insistence of sociologists that both truth and error are equally amenable to sociological analysis'. (Woolgar 1993, p. 42)

In what way were the social constructivists misinterpeting the philosophers? Simply this: Lakatos and the others were not claiming that belief in true ideas is uncaused. Rather, they were suggesting (perhaps in too subtle a way for the social constructivists to comprehend) that true beliefs are explained (given the *ceteris paribus* background assumption that the scientists concerned have normal neural mechanisms and so forth) by the fact that there are reasons, warrants, evidence and so forth available that indicate that the ideas *are indeed true*. In other words, the content of ideas, and the *content* of the justificatory arguments that support acceptance of these ideas as being true, *are part of the causal story* that needs to be told (and this is an *internal* story)! The philosopher Larry Laudan put this quite powerfully when he wrote a critique of Bloor's foundational book:

> If Bloor's caricatures were to be accepted, we should believe that most philosophers ... have maintained that there is literally nothing that causes us to believe what is true

and that nothing is causally responsible for rational action and rational belief. But Bloor's analysis of the philosophical tradition will not stand up to scrutiny. For as long as we know anything about the history of philosophy, epistemologists have been concerned to explain how to discover the true and the rational. The suggestion that most philosophers have believed that true beliefs just happen, that rational behavior is uncaused, that only 'aberrant' belief is part of nature's causal nexus, is hard to take seriously. (Laudan 1981, p. 178)

I should note in passing that social constructivists (not only the radicals, but the moderates as well) also make the point that the internal/external distinction breaks down once we recognize that the work of scientists goes on in a socio-cultural setting that influences it; but it should be clear (although it is not clear to many constructivists) that it does not follow from this that the point made by Lakatos and others can be disregarded. For even if the distiction cannot be maintained (my own view is that it is a rough or permeable distinction rather than a watertight one), what were previously called 'internal reasons' still do not go away – these epistemic or non-social factors (as Kornblith calls them) remain part of the total amalgam of internal/external factors, whatever language we use to describe this. But, for the radical social constructivists, these internal/non-social factors disappear from the account altogether, as we have seen in passages from Bloor, Woolgar, and Latour. Hilary Kornblith has described a moderate position which recognizes the role played by both internal and external factors (however they are grouped together or interact):

> On any account, then, belief acquisition and retention must be seen as a product of both social and non-social factors, and both kinds of factors will come in for investigation in the course of epistemic evaluation. Moreover, on any reasonable account, both kind of factors will play a role in explaining individual differences in belief. There is still room for a great deal of disagreement about the relative weight of these two kinds of factors.(Kornblith 1994, p. 102)

It is time to bring this over-long discussion to an end. But in concluding it is worth highlighting what, according to my analysis, is the central issue at stake in considering the 'causes of belief' and the position of many of the social constructivists on this matter – a position that has important implications for education, as Slezak has noted (Slezak 1994, Part 1): The adherents of the strong programme, and many of those engaged in the social studies of knowledge together with fellow-travellers from the postmodernist camp, have a deficient model of what it is to be human (although I think that many of them would be surprised to learn this). They see humans as *behaving*, as being objects which are caused to do certain things because of the publicly observable 'external' forces (sociological, political, economic, and so forth) that bear down upon them, and perhaps because of psychological drives (or 'internal forces') such as the lust for power or fame. As Slezak has noted in an important discussion,

> the externalist conception of theories as caused by features of the social milieu is a form of stimulus control theory akin to Skinnerian behaviourism. On this conception, beliefs

are the causal consequence of external environmental, social factors rather than internal, cognitive, intellectual ones. (Slezak 1994, Part 1, p. 267)

Philosophers of science and epistemologists generally hold what I would argue is a more defensible model, one which has been widely canvassed by recent philosophers of social science in the English-speaking world and for over a century by hermeneuticists and philosophers from the Continent: Humans do not merely behave as a result of external or internal forces (although undeniably these are of importance), but they also engage in *voluntary actions* as a result of the ideas, beliefs, meanings, motives, and knowledge that they possess (see Phillips 1992, ch. 1, for further discussion of, and references concerning, this model). This is not to say that there is no interaction between the two domains of external and internal forces on the one hand, and voluntary action on the other; clearly this is often quite substantial. But to treat knowledge construction as a behavior rather than as an action is, in a sense, to dehumanize it and to refuse to take it seriously as knowledge (that is, as something that is believed because of good although not unassailable reasons, even if these reasons are not – and could never be – immune from social influences).

ACKNOWLEDGEMENTS

This paper is based on material I am collecting for a book *The Many Faces of Constructivism*; the current essay is a shorter and (hopefully) strengthened version of a paper I presented to the 'Symbolic Systems Seminar in Education' at Stanford University in 1995, and at the annual meeting of the Philosophy of Education Society of Australasia, Melbourne, December 1995. I am indebted to participants for the helpful comments I received on both of those occasions.

NOTE ADDED IN PROOF

As this paper was about to go to press, there was another "cause celebre": Physicist Alan Sokal had a spoof paper on the social constructivist implications of quantum gravity theory accepted by *Social Text*, a "cultural studies" journal with which Stanley Fish has links; the day the paper appeared, Sokal revealed the joke in another journal, *Lingua Franca*. Fish, doing a good job in covering his embarrassment, responded with a strongly worded op-ed piece in the *New York Times* (Fish 1996). And so the disputes rage on.

REFERENCES

Barnes, B.: 1994, 'How Not to do the Sociology of Knowledge', in Allen Megill (ed.), *Rethinking Objectivity*, Duke University Press, Durham N.C.

Barnes, B. & Bloor, David: 1982, 'Relativism, Rationalism and the Sociology of Knowledge', in M. Hollis and S. Lukes (eds.), *Rationality and Relativism*, MIT Press, Cambridge, Mass.
Bloor, D.: 1976, *Knowledge and Social Imagery*, Routledge, London.
Bloor, D.: 1981, 'The Strengths of the Strong Programme', *Philosophy of the Social Sciences* **11**(2), 199-213.
Brown, J.R. (ed).: 1984, *The Rationality Debates: The Sociological Turn*, Reidel, Dordrecht, The Netherlands.
Cherryholmes, C.: 1988, *Power and Criticism*, Teachers College Press, New York.
Cole, S.: 1992, *Making Science*, Harvard University Press, Cambridge, Mass.
Collins, H.: 1981, 'Stages in the Empirical Program of Relativism', *Social Studies of Science* **11**, 3-10.
Collins, H.: 1992, *Changing Order*, University of Chicago Press, Chicago.
Fish, S.: 1996, 'Professor Sokal's Bad Joke', Op-Ed. *New York Times*, May 21.
Forman, P.: 1995, 'Truth and Objectivity, Part 1', *Science* **269**, July 28, 565-567.
Fuller, S.: 1988, *Social Epistemology*, Indiana University Press, Bloomington, Indiana.
Fuller, S.: 1993, *Philosophy of Science and its Discontents*, Guilford Press, New York.
Goldman, A.: 1992, *Liaisons*, MIT Press/Bradford, Cambridge, Mass.
Gross, P. & Levitt, N.: 1994, *Higher Superstition*, Johns Hopkins University Press, Baltimore.
Hawkesworth, M.: 1989, 'Knowers, Knowing, Known: Feminist Theory and Claims of Truth', *Signs* **14**(3), 533-557.
Hollis, M.: 1982, 'The Social Destruction of Reality', in M. Hollis and S. Lukes (eds.), *Rationality and Relativism*, MIT Press, Cambridge, Mass.
Keller, E.F.: 1995, 'Science and Its Critics', *Academe* **81**(5), 10-15.
Kornblith, H.: 1994, 'A Conservative Approach to Social Epistemology', in F. Schmitt (ed.), *Socializing Epistemology: The Social Dimensions of Knowledge*, Rowman and Littlefield, Lanham, Md.
Kuhn, T.S.: 1992, *The Trouble with the Historical Philosophy of Science*, The Robert and Maureen Rothschild Distinguished Lecture, Harvard Department of the History of Science, Cambridge, Mass.
Lakatos, I.: 1978, 'History of Science and Its Rational Reconstructions', in I. Lakatos, *The Methodology of Scientific Research Programs*, Cambridge University Press, Cambridge.
Latour, B.: 1992, 'One More Turn After the Social Turn', in E. McMullin (ed.), *The Social Dimensions of Science*, University of Notre Dame Press, Notre Dame, Indiana.
Latour, B. & Woolgar, S.: 1986, *Laboratory Life: The Construction of Scientific Facts*, Princeton University Press, Princeton, N.J.
Laudan, L.: 1981, 'The Pseudo-science of Science?', *Philosophy of the Social Sciences* **11**, 173-198.
Laudan, L.: 1990, *Science and Relativism*, University of Chicago Press, Chicago.
Lovie, S.: 1995, 'Review Note: Stephen Cole, Making Science', *Theory and Psychology* **5**(4), 611-612.
Lyotard, J.-F.: 1984, *The Postmodern Condition: A Report on Knowledge*, Manchester University Press, Manchester.
Megill, A. (ed.): 1994, *Rethinking Objectivity*, Duke University Press, Durham, N.C.
Newton-Smith, W.H.: 1981, *The Rationality of Science*, Routledge, Boston.
New York Review of Books: 1995, 'Feminism and Philosophy: An Exchange', **xlii**(6), 48-49.
Nussbaum, M.: 1994, 'Feminists and Philosophy', *New York Review of Books*, **xli**(17), Oct. 20, 59-63.
Phillips, D.C.: 1987, *Philosophy, Science, and Social Inquiry*, Pergamon Press, Oxford.
Phillips, D.C.: 1992, *The Social Scientist's Bestiary*, Pergamon Press, Oxford.
Phillips, D.C.: 1995, 'The Good, the Bad, and the Ugly: The Many Faces of Constructivism', *Educational Researcher* **24**(7), October, 5-12.

Rorty, R.: 1979, *Philosophy and the Mirror of Nature*, Princeton University Press, Princeton, N.J.
Russell, B.: 1948, *An Outline of Philosophy*, Allen and Unwin, London.
Scheffler, I.: 1967, *Science and Subjectivity*, Bobbs-Merrill, New York.
Slezak, P.: 1994, 'Sociology of Scientific Knowledge and Scientific Education, Part 1', *Science & Education* **3**, 265-294.
Slezak, P.: 1994, 'Sociology of Scientific Knowledge and Science Education, Part 2: Laboratory Life Under the Microscope', *Science & Education* **3**, 329-355.
Shapin, S.: 1982, 'History of Science and Its Sociological Reconstructions', *History of Science* **xx**, 157-211.
Sismondo, S.: 1996, *Science Without Myth*, State University of New York, Albany, New York.
Times Higher Education Supplement: 1994, 'Sociology Row Erupts at BA', Sept. 16, p. 44.
Woolgar, S.: 1993, *Science: The Very Idea*, Routledge, New York.

Sociology of Scientific Knowledge and Scientific Education: Part I

PETER SLEZAK

School of Science and Technology Studies, University of New South Wales, Kensington, NSW 2033, Australia

ABSTRACT: This article is the first of two that will examine the claims of contemporary sociology of scientific knowledge (SSK) and the bearing of these claims upon the rationale and practice of science teaching. It is maintained that if the claims of SSK are true then there are serious, and educationally and culturally deleterious, implications which follow. The two articles will argue that, fortunately, the claims of SSK for the external causation of scientific belief are baseless. And thus science teachers should resist admonitions to accept the findings of the sociology of science.

> 'I look forward to the day when the last proponent of the 'strong program' in the sociology of science is strangled in the entrails of the last expert in the theory of metaphor'
> Alasdair MacIntyre, 1988

EDUCATION OR INDOCTRINATION? IDEAS OR IDEOLOGY?

Bertrand Russell suggested that a wise system of education - one which considered the interests of the students and the society - would not aim at instilling allegiance to any particular view or particular party, but rather, it would aim at enabling them to choose intelligently between views and parties. Such an education 'would aim at making them able to think, not at making them think what their teachers think' (1961, 401–2).

The same ideal is at the heart of modern science itself as one of its constitutive principles. As Popper (1963), Guthrie (1962) and other scholars have noted, the origins of science among the Presocratics had its characteristic novelty, not in the ideas themselves which might have become merely a different orthodoxy to replace the Homeric gods on Olympus; rather, the novelty was to be seen in the tradition of critical inquiry - the demand to improve upon the teacher's story rather than merely to perpetuate it unquestioningly. In this sense, science has a special importance in the curriculum through the values it embodies and, correspondingly, the history of science is more than a catalogue of past achievements, but an exemplification of these values.

Russell's pedagogical precept is, however, implicitly challenged today by certain doctrines which are gaining a considerable following in our universities and among educationalists. The doctrines of social constructivism take scientific theories to reflect the social milieu in which they emerge and, rather than being founded on logic, reason and evidence, beliefs are taken to be causal effects of the prevailing context. Thus, typical of social constructionists, Latour and Woolgar (1979, 1986) for example, consider scientific success to be merely the ability of a theory's proponents to 'extract compliance' from others. What is usually called the 'plausibility' or warrant for beliefs is not, in fact, an intellectual or cognitive question, but only a matter of the 'balance of forces' and political allegiances.

The educational implications of their doctrines have not been explicitly drawn out or even addressed by proponents of recent social studies of science, but these implications are not difficult to discern. By explicitly repudiating the role of rational considerations, such views entail that Russell's educational ideals and the values embodied in the Western intellectual tradition of science must be fundamentally misguided and unrealizable in principle. That is, if beliefs are intrinsically the products of 'external' factors such as social causes and interests rather than 'internal' considerations of evidence and reason, then it is an illusion to imagine that education might serve to instil the capacity for critical thought which Russell recommends. On these views, the very distinction between education and indoctrination becomes otiose; ideas are merely ideology, and pedagogy is merely propaganda.

For example, the relativism of social constructivist theories makes it impossible for teachers to offer the obvious objections to such excesses as to be seen in the notorious Lysenko era in Soviet biology. In particular, one cannot complain that Lysenkoism was a perversion of scientific truth. The purges of orthodox geneticists must count as an instance of successful science according to the criteria of Latour and Woolgar. Or, to take another example having urgent contemporary relevance for science education, one's approach to the issue of 'equal time' in the classroom for 'Creation Science' will depend crucially on the stance taken regarding the doctrines of the sociology of scientific knowledge. Specifically, the distinction between science and pseudo-science takes on a very different complexion for a social constructivist (see Pinch and Collins 1984). From this point of view, such questions cannot even arise if understood as questions about the relative merits of the competing theories. The very notion of merit, as indeed the honorific label of 'Science' itself, is repudiated by social constructivist theories as reflecting only interests and power relations among different groups. Accordingly, social constructivists cannot argue against the introduction of Creation Science into school biology curricula by invoking the usual cognitive criteria. Again, the political success of fundamentalism would *ipso facto* constitute scientific success and one cannot complain against their educational ambitions on the grounds of scientific disreputability. Likewise, if Pinch (1993) and Ashmore (1993) are to be taken seriously in their effort to 'alter the grounds of consensus', then the science curriculum should include Blondlot's notorious 'discovery' of N-Rays alongside that of Roentgen's X-rays without dismissing the former as a spurious and discredited episode in the history of science.

Whatever the ultimate judgement of its merits, there can be no doubt about the radical, iconoclastic nature of the claims of the sociology of scientific knowledge (SSK). The self-advertising, at least, proclaims a most fundamental revision of our conception of science and its claims to reliable knowledge of the world. SSK claims to have unmasked the pretensions of science. Instead of deserving a privileged status and an honorific label, on this view, science and its knowledge claims are to be relegated to the same category as all other merely social practices – perhaps like art, law and dress fashions. The implications for science education are correspondingly far-reaching, as I will indicate presently.

Quite apart from the question of *what* should be taught, there is the further educational issue of *how* science should be taught. Traditional conceptions of both the ends and the means of a science education are gravely challenged if social constructivist doctrines are taken seriously. If, as Bloor (1976) suggests, beliefs are not a matter of reasons, evidence and other rational considerations, then teaching cannot involve conveying ideas through *understanding*. We will see that the externalist conception of theories as caused by features of the social milieu is a form of stimulus control theory akin to Skinnerian behaviourism. On this conception, beliefs are the causal consequence of external environmental, social factors rather than internal, cognitive, intellectual ones. If taken seriously, this entails that the actual contents or 'ideas' of science are merely an epiphenomenal by-product of the social processes. The attack on 'psychologism' by proponents of the social studies of science is an attack on the idea that beliefs are acquired and justified through mental reasoning processes. The educational consequences could not be more dramatic: the assumption that teaching involves the inculcation of ideas and understanding based on considerations of evidence and argument must be an illusion. For example, a teacher's explanation which purports to reveal the logical connections among ideas and their warrant must be misguided about the grounds of belief or, indeed, about the notion that there might be *grounds* of belief at all, in principle. If beliefs are the products of social contexts rather than intellectual appraisal, then 'learning' so-called must be mere conditioning – the arbitrary pairing of 'ideas' with reinforcing social stimuli. Correspondingly, on this view teaching cannot be a matter of producing understanding or insight but only compliance.

Accordingly, this article offers a critical survey of some key ideas and developments in recent social studies of scientific knowledge. The discussion here makes no claim to being a representative or comprehensive survey of the field since such accounts are readily available elsewhere. Instead, my discussion seeks to present a selective, critical analysis of some of the key ideas of the field which brings into relief their essential and problematic features. Particular attention is devoted to the work of David Bloor, Bruno Latour and Steve Woolgar whose writings are widely seen as among the foundational classics of the field.

It must be noted at the outset that the field has been characterised by polemics of an unusually acrimonious sort. For example, Mario Bunge (1991) has described most of the work in the field as 'a grotesque cartoon of scientific research'. In a similar vein, the philosopher David Stove (1991) has recently written of these doctrines as 'philosophical folly' and as 'a stupid and discreditable business' whose authors are 'beneath philosophical notice and unlikely to benefit from it'. In his scathing remarks, Stove describes such ideas as an illustration of the 'fatal affliction' and 'corruption of thought' in which people say things which are bizarre and which even they must know to be false. Laudan, who has been among the few philosophers to make systematic critical analyses of social constructivism has recently (Laudan 1990, p.x) characterized this 'rampant relativism' as 'the most prominent and pernicious manifestation of anti-intellectualism in our time'. From such reactions it should be clear that it is difficult to give a neutral exposition of the subject and, therefore, it is important to emphasize the partisan nature of the present account which shares the critical standpoint of the views just noted.

THE POVERTY OF CONSTRUCTIVISM[1]

Although the recent proponents of social constructivism make little effort to trace their own intellectual antecedents, there can be little doubt about the affinities of their doctrines with the Hegelian historicism which Popper so bitterly denounced as 'this despicable perversion of everything that is decent'. For SSK, as for Hegel, 'History is our judge. Since History and Providence have brought the existing powers into being, their might must be right . . .' (Popper, 1966a, p.49). The unmistakable parallel is seen in their essentially similar answers to Popper's fundamental question 'who is to judge what is, and what is not objective truth?' He reports Hegel's reply that 'The state has, in general . . . to make up its own mind concerning what is to be considered as objective truth' and adds 'With this reply, freedom of thought, and the claims of science to set its own standards, give way, finally, to their opposites' (1966a, p. 43). Though Hegel's doctrines are expressed in terms of the 'State', the essential idea is that political success is *ipso facto* the criterion of truth. We will see precisely this idea resuscitated in Latour and Woolgar, Pinch and Collins, and the entire enterprise of contemporary sociology of scientific knowledge. This is a historical relativism according to which truth is dependent on the *zeitgeist* or spirit of the age, and it is the one which Popper charges with helping to destroy the tradition of respecting the truth (see Popper 1966a, p. 308 fn 30).[2]

David Bloor (1976) begins his manifesto for the Edinburgh School Strong Programme by specifically addressing and attempting to rebut Popper's critique of historicism – thereby acknowledging both the historicism of his own programme and the relevance of Popper's critique.[3]

MERTONIAN NORMS: ETHOS OF SCIENCE & INSTITUTIONAL IMPERATIVES

Thus, not only fundamental conceptions of the content of science are at stake. No less serious than the consequences for traditional views of scientific theories are the consequences for ethical principles and their place in any science curriculum. Besides the facts and theories conveyed in a science education are certain values and norms of conduct. Some of these are more specifically pertinent to the practice of science, while others are general moral precepts of the community at large. Besides the academic conventions concerning citations, acknowledgments and other scholarly practices are the noble ideals of objectivity and truth which have been seen as among the important human values embodied in the scientific outlook. The inculcation of these broader values has been widely taken to be among the important functions of a science education, but the doctrines of social constructivism may be seen as posing a fundamental challenge to this ethical dimension of science education as well.

Pinch and Collins (1984), for example, endorse the idea that the facts of science are 'socially constructed' through 'negotiation' and other political acts. Accordingly, they maintain a scrupulous neutrality concerning the substantive merits of different theories such as those of orthodox science compared with astrology and parapsychology. Correspondingly, they have no grounds to distinguish honourable scientists from unscrupulous charlatans. They see only the external manoeuvres, rhetoric and tactics used by each side in their disputes. One could hardly get a clearer account of the normative implications which follow from the social constructivist approach – though these implications are not actually made explicit: Deception is just the way the game is played and, consequently, in one instance, Pinch and Collins suggest that, if the scientists had managed to act more dishonestly to cover up contrary evidence, they would have been more successful; that is, 'they could have maintained their position'. This is the only criterion available for assessment to Pinch and Collins. Since the 'facticity' of any claim is just a matter of how the claims are presented, there can be no conception of truth or honesty in this view of science. Indeed, when the scientists admit their earlier lapse of concealing contrary evidence, they are ridiculed by Pinch and Collins for their grandiose, mythical pretensions and for clinging to a naive understanding of scientific method: They have failed to appreciate the way that facts are socially constructed, negotiated and constituted entirely by their manner of presentation. That is, the literature can either construct or 'dissolve the facticity of the claims'. There are no facts as such, only 'facticity' which is entirely a matter of public presentation. For the enterprise of Pinch and Collins, the exposure of duplicity is not essentially different from the deconstruction of ordinary scientific practice. If we drop the jargon, their point is simply that truth is what you can get away with.

Pinch and Collins are not exceptional in embracing such views. Latour and Woolgar espouse essentially the same ideas by attempting to eradicate any distinction between truth and falsehood. They profess to reject the belief in 'the intrinsic existence of accurate and fictitious accounts per se' (1979, p. 284). On their view, too, all of science is merely the 'construction of fictions'. Accordingly, on this account too it is impossible to distinguish fairness from fraud since both are ways of constructing fictions. For these sociologists the success of any theory is entirely a matter of 'increasing the number of people from whom it extracts compliance'. On this theory a repressive totalitarian regime must count as a paradigm of scientific progress. The usual ethical distinctions only make sense on the assumption of certain cognitive standards of judgement and on the assumption that there is an intrinsic difference between the knowledge claims towards which behaviour might be directed. In the absence of such categories, the scientist like Cyril Burt who fraudulently manufactures his evidence cannot be meaningfully distinguished from the honest researcher whose data are also 'constructed', albeit in different ways. Latour and Woolgar assert 'Each text, laboratory, author and discipline strives to establish a world in which its own interpretation is made more likely by virtue of the increasing number of people from whom it extracts compliance' (1986, p. 285).

The problem arises from the social constructivist rejection of the famous Mertonian (1942) norms of universalism, communism, disinterestedness and organized skepticism which constitute the 'ethos of science'. Merton described these as institutional imperatives, being 'moral as well as technical prescriptions' – 'that affectively toned complex of values and norms which is held to be binding' on the scientist. As Merton observes, these institutional values are transmitted by precept and example, presumably in the course of the scientist's education. It is difficult to see how someone committed to the social constructivist view can either teach or conduct science according to the usual rules in which truth, honesty and other measures of worth are taken seriously. The tension between one's theory and one's practice is obvious: either the constructivist acts according to the usual institutional norms in the teaching and practice of science and is thereby in glaring contradiction with constructivist theory, or the teacher-scientist can be consistent and thereby unwilling to recognise the values of truth, honesty and integrity. Although it seems both important and unavoidable, I am not aware that the sociologists have either posed or answered this question directly.

As already indicated, for educators the grounds for concern are seen clearly enough by reflecting on the fact that the sociology of scientific knowledge could have offered no principled objection to teaching the racial theories of *Mein Kampf* when they were believed by a majority. The pernicious consequences for education and, consequently, for social life, arise from the systematic 'externalist' pretence that the conceptual, intellectual content of ideas has no intrinsic meaning or essential relation to anything outside itself. The meaning of a text is indeterminate and infinitely negotiable and, therefore, to be explained by social factors rather than cognitive, rational ones. This is a version of the notorious Deconstructionist affectation '*il n'y a pas de hors texte*'. On such an account science becomes a play with signs having no significance other than one attributed through social negotiation and construction. A leading constructivist, Andrew Pickering (1992, p. 22), has recently asserted that no less than 'The foundations of modern thought are at stake here'. On this assessment, at least, all sides may readily agree.

SOCIOLOGY OF SCIENTIFIC KNOWLEDGE: THE EDINBURGH STRONG PROGRAMME

Measured by external criteria, recent developments in the Sociology of Scientific Knowledge can boast considerable success. The publication of books and journals, the establishment of departments and the growth of professional societies all attest to the apparent vigour of the academic enterprise. The republication in new editions of two of the foundational 'classics' of the field (Bloor 1991b, Latour and Woolgar 1986) is further evidence of the continuing interest and importance of the field. Moreover, the relatively new discipline gives every appearance of making progress with ever new theoretical developments amid vigorous controversies about its fundamental principles and methods. One of its 'schools' declares itself to be the third identifiable phase in an evolutionary process which has taken the social studies of science to deeper and deeper insights (Woolgar 1988).

David Bloor has the distinction of having written the small book *Knowledge and Social Imagery* (1976) which launched the so-called 'Edinburgh Strong Programme' in the sociology of scientific knowledge. The undoubted appeal of Bloor's work to many was its iconoclastic approach to old-fashioned theories. Bloor was self-consciously heralding a radical enterprise intended to discomfit traditionalists in philosophy and sociology. Foreshadowing the provocation of later works, Bloor's preface to the first edition of his book already hints darkly that the inevitable resistance by philosophers and sociologists to his doctrines will be due to uncomfortable secrets that they would wish to hide. Bloor asserts that his approach to science from a sociological point of view encounters resistance because 'some nerve has been touched'. He announces his bold intention to 'despoil academic boundaries' which 'contrive to keep some things well hidden' (1976, p. ix).

Bloor was right about some nerve having been touched, though he may have misdiagnosed the nature and the source of the noxious irritation. He devotes a chapter of his book to a kind of psychoanalysis of his opponents by speculating about the 'sources of resistance' to the Strong Programme which he attributes to hidden, indeed primitive, motives involving the fear of sociology's desacralizing of Science and its mysteries. One might suggest alternative reasons for the resistance to such sociological analyses, but

Bloor sees only repressed impulses concerning the 'sacred' and the 'profane' leading to 'a superstitious desire to avoid treating knowledge naturalistically' (1976, p. 73). Bloor imagines that the 'threatening' nature of any investigation into science itself has been the cause of a 'positive disinclination to examine the nature of knowledge in a candid and scientific way' (1976, p. 42). However, this disinclination to examine knowledge, and the need to keep it mystified through fear of desecration, is difficult to reconcile with the fact that every philosopher since Plato has been centrally concerned with the problem of knowledge and its justification.

THE SOCIAL CONSTRUCTION OF SOCIAL CONSTRUCTIVISM

In his manifesto, Bloor (1976, p. 3) had declared that the central claims of the Strong Programme he launched were 'beyond dispute', and Barnes (1981, p. 481) begins an article asserting that in the short time since its advent 'developments have occurred with breathtaking speed' and 'the view that scientific culture is constructed like any other is now well elaborated and exemplified'. Indeed, Barnes goes so far as to say that it has passed from having first been denounced as absurd, to having been disdained as true and 'now it is recognized as so important that its opponents are starting to say that they themselves discovered it'. Barnes neglects to mention which opponents now wish to claim credit for discovering these doctrines, nor does he acknowledge that some still wish to denounce them as absurd.

It is hard to imagine researchers in genuinely well established disciplines such as physics or biology making such frequent prefatory assertions about the security of their achievements. This level of self-congratulatory hyperbole has prompted Thomas Gieryn (1982, p. 280) to comment upon these 'defences and re-affirmations' as 'expressions of hubris' and 'exaggerations passing as fact'. Gieryn (1982, p. 293) has suggested that the radical findings of the new sociology of science 'are "new" only in a fictionalized reading of antecedent work'. In particular, Robert Merton's article on 'The Sociology of Knowledge' (1957) had specifically enunciated the very central doctrine of the Strong Programme. Merton wrote:

> The 'Copernican revolution' in this area of inquiry consisted in the hypothesis that not only error or illusion or unauthenticated belief but also the discovery of truth was socially (historically) conditioned... The sociology of knowledge came into being with the signal hypothesis that even truths were to be held socially accountable, were to be related to the historical society in which they emerged. (1957, p. 459)

As we will see presently, this statement articulates the core of Bloor's Strong Programme, particularly the tenets of causality, impartiality and symmetry. Merton's careful analytical framework provides a systematic basis for comparing what he described as 'the welter of studies which have appeared in this field' and it remains a most valuable document for the same reason today. In particular, Merton carefully distinguishes the various possible claims and theses which have been asserted in the more recent literature.[4]

CONTEXTS, CONTENTS AND CAUSES.

Bloor states the four tenets of his sociology of scientific knowledge as follows:
1. It would be causal, that is, concerned with the conditions which bring about belief or states of knowledge.
2. It would be impartial with respect to truth and falsity, rationality or irrationality, success or failure. Both sides of these dichotomies will require explanation.
3. It would be symmetrical in its style of explanation. The same types of cause would explain, say, true and false beliefs.
4. It would be reflexive. In principle its patterns of explanation would have to be applicable to sociology itself.

Though it had appeared in different guises before, such as in the historicism of Hegel and Marx, the radical idea at the heart of the Strong Programme was to go beyond the sociological studies which stopped short of considering the actual substantive content, the ideas, of scientific theories as an appropriate domain for sociological investigation. Previously, sociological studies paid attention only to such things as institutional politics, citation patterns and other such peripheral social phenomena surrounding the production of science, but had not ventured to explain the cognitive contents of theories in sociological terms. Since this crucial point has been obscured, its importance for appreciating subsequent developments cannot be overstated. The opening sentence of Bloor's book asks 'Can the sociology of knowledge investigate and explain the very content and nature of scientific knowledge?' (1976, p. 1) – that is, 'knowledge as such, as distinct from the circumstances of production'.

The alleged failure of previous sociological studies to touch on the contents of scientific belief was portrayed by Bloor (1976, p. 8) as a loss of nerve and a failure to be consistent. Karl Mannheim, among the founders of the sociology of knowledge, is characterised as failing to make the logical extension of his approach from knowledge of society to the knowledge of nature as well. The epistemological pretensions of the Strong Programme – its relativist challenge – derives from this thorough-going application of the sociological principle which seeks to explain the hitherto exempted knowledge claims. The ambitions of Bloor's program are explicit from the outset, for he complains that previous sociologists, in 'a betrayal of their disciplinary standpoint' had failed to 'expand and generalise' their claims to all knowledge: '... the sociology of knowledge might well have pressed more strongly into the area currently occupied by philosophers, who have been allowed to take upon themselves the task of defining the nature of knowledge' (1976, p. 1).[5]

IMPARTIALITY

'Sociology is only for deviants' (W.H. Newton-Smith)

Once sociological explanations are sought for the contents of theories and not only the circumstances of their production, it was natural to make the extension to all fields of knowledge in the manner just noted. However, there remains some further scope for extension of the sociological approach. Merton, like Mannheim, argued that the explanations for ideas may be sought in social factors only if they cannot be shown to be 'immanently' or rationally determined. That is, theories judged to be correct and founded on rational considerations are not in need of sociological explanation in the way that false and irrational theories are. In keeping with common practice, manifestly absurd beliefs are taken to require some special accounting of a kind not sought in other cases. In this sense, traditional conceptions relegated sociology to the dross of science, to its residue of false and irrational beliefs. Bloor was explicitly rescuing sociology from this ignominious role by asserting the appropriateness of sociological explanations for all of science regardless of evaluative judgements such as truth and falsity, rationality and irrationality, success or failure. Indeed, in what J.R. Brown (1989, p. 7) describes as 'the very antithesis of the sentiments embodied in the recent sociological turn', Laudan has reasserted the earlier asymmetrical methodological approach in what he calls the 'Arationality Principle', suggesting that 'the sociology of knowledge may step in to explain beliefs if and only if those beliefs cannot be explained in terms of their rational merits' (Laudan 1977, p. 202). Speaking for the Impartiality Principle, Barnes says that 'What matters is that we recognise the *sociological* equivalence of different knowledge claims . . . as a methodological principle we must not allow our evaluation of beliefs to determine which form of sociological account we put forward to explain them' (1977, p. 25).

SYMMETRY & NEUTRALITY (BLONDLOT WAS FRAMED)

Perhaps most telling is the attitude of SSK to charlatanism, fraud and discredited science. As noted, the tenets of 'symmetry' and 'impartiality' have meant that sociological studies professed to maintain a scrupulous indifference to the truth, rationality, warrant or intellectual merits of scientific theories on the grounds that sociological causes must be sought regardless of any evaluation. In fact, this feigned suspension of judgement involves considerable *mauvais fois* – a blindness which is far from neutral on questions of merit. In practice it has meant a tacit advocacy of discredited or disreputable science. Implicit in the conception of a hegemonic orthodoxy is the idea that rejected theories are only politically rather than intellectually disadvantaged. Latour and Woolgar explicitly portray their own theories as merely lacking popularity rather than plausibility.

Thus, most recently, having to confront the manifest disingenuousness of the 'symmetry' stance, Pinch (1993) and Ashmore (1993) go so far as to defend the supposed merits of unorthodox and rejected theories. Not least, this policy is evidently taken to include the case of fraud since this 'is to be seen as an attributed category, something made in a particular context which may become unmade later' (Pinch 1993, p. 368). Ashmore proposes a radical skepticism concerning the exposé of notorious cases of misguided science such as that of Blondlot's spurious N-rays. Amid the usual jargon-laden pseudo-technicality, such an approach amounts to actually promoting the alleged scientific 'merits' of such discredited cases. Pinch writes of 'making plausible the rejected view' (1993, p. 371) and Ashmore is perfectly explicit: 'To put it very starkly, I am looking for justice! ... in a rhetorically self-conscious effort to alter the grounds of consensus' (1993, p. 71). Thus, where proponents of SSK have displayed an undisguised prejudice in favour of unorthodox or discredited theories, Pinch pretends to defend their 'symmetry' which he spuriously distinguishes from 'neutrality'. On Pinch's account it seems that 'symmetry' has come to mean something like affirmative action for disreputable theories. Presumably, it follows that Blondlot should be given equal time with Roentgen in physics lessons.

DIAMETRICALLY OPPOSED 'TELEOLOGICAL' VIEW AS ACAUSAL

Bloor has construed the asymmetrical treatment of earlier views as a failure to seek or ascribe causes to true, rational beliefs. In his *Knowledge and Social Imagery* (1976), Bloor characterized the 'autonomy' view he is opposing:

> One important set of objections to the sociology of knowledge derives from the conviction that some beliefs *do not stand in need of any explanation, or do not stand in need of a causal explanation*. This feeling is particularly strong when the beliefs in question are taken to be true, rational, scientific or objective. (1976, p. 5; emphasis added)

Elsewhere Bloor characterizes the opposing view as 'the claim that nothing makes people do things that are correct but something does make, or cause, them to go wrong' and that in the case of true beliefs causes do not need to be invoked' (1976, p. 6). From these and other such remarks it is perfectly clear that Bloor intends to make an absolute distinction between the 'teleological' view which inclines its proponents to 'reject causality' (1976, p.10) and '*the* causal view' – that is, the sociological approach of the Strong Programme. On Bloor's own account, the viability of the strong programme rests on the tenability of this dichotomy and, in particular, the falsity of the 'teleological model'. There could be no more crucial issue for the strong programme. As we will see presently, not only such programmatic statements are at stake, since the substantive research program of case studies purports to establish the causal claim in the form of the connections between contingent historical circumstances and scientific theories. Laudan (1981, p. 178) has characterized Bloor's acausal attribution to philosophers as an absurd view which cannot plausibly be attributed to any philosopher at all. However, in a remarkable passage, Bloor has responded to Laudan by attempting to deny these patent and quite explicit earlier intentions. The point is of considerable importance, for as we have seen, the entire edifice of the Strong Programme rests on this claimed opposition – which is, however, completely spurious. Bloor's discomfort is perhaps understandable, and his tactics here and subsequently are perhaps the only alternative to retraction and abandoning the enterprise altogether. Bloor concedes, after all, that no philosopher ever held the a-causal view, but somewhat implausibly protests that he was arguing for a different, if somewhat obscure, alternative thesis. Quite aside from being a reversal, Bloor's concession here is potentially more damaging than Bloor acknowledges. Once it is seen that the 'teleological' position is not the straw-man Bloor represented it to be, then its implications must be confronted seriously. The need to resort to the lame responses we have just seen in the face of the most serious criticism is a clear sign of the bankruptcy of the sociological enterprise in spite of Bloor's attempts at camouflage. If the foregoing evidence were not clear enough, Bloor has now left no doubt about the analysis and its conclusions. In the new second edition of his book, in the crucial section on the 'Autonomy of Knowledge' dealing with the problem of causation, we discover certain judicious changes to the original text whose rationale is clearly to avoid the charges made by Laudan.[6]

Bloor's discomfort is understandable since his statement of the conditions under which the programme retains its plausibility left no room for compromise and no way out. Bloor had declared forthrightly:

> There is no doubt that if the teleological model is true then the strong programme is false. The teleological and causal models, then, represent programmatic alternatives which quite exclude one another. (1976, p.9)

BORN-AGAIN BEHAVIOURISM

As noted earlier, in Bloor's work, the claim of 'existential determination' of knowledge takes the form of a purported causal connection between the ideas or theory contents and the social milieu. This causal claim is actually a version of stimulus-control theory akin to that of Skinnerian Behaviourism and, not surprisingly, in his later work Bloor (1983) explicitly endorses such notorious theories. It is this causal claim which is among the central, distinctive tenets of Bloor's Strong Programme and the basis for contrasting it with alternative allegedly a-causal 'teleological' approaches. In characterising opposing rationalist or 'teleological' views, quoting Wittgenstein, Bloor (1983, p. 6) refers to explanations which postulate mental states as infected by the 'disease' of 'psychologism'. Bloor's frontal assault on mental states as having explanatory force is an intrinsic part of the defence of the radically alternative sociological approach to explaining science but this bold stance left his programme vulnerable to a case on the other side whose strength he had grievously underestimated. For example, anachronistically Bloor's programme depends on rejecting the reality of mental states such as images. However, this position is thirty years and a major scientific revolution too late.

The recidivism and bankruptcy of Bloor's programme has been disguised more recently only by attempts to water down its original extreme radicalism by claiming its compatibility with individual psychology. This tactic suggests that a retrograde amnesia appears to have afflicted Bloor and his colleagues who now write as if there were no extravagant claims and no presumptuous propaganda on which the entire movement was built (See Bloor's 1991a response to Nola 1991 and new Postscript, 1991b).

BLOOR'S BELATED BLANDNESS

Bloor's (1991b) recent protests that his views are entirely consistent with cognitive science cannot be taken seriously. It can be asserted at all only because Bloor now pretends that the sociological thesis at stake is merely whether or not there are social aspects to science. But this weak and uncontroversial thesis is not the original doctrine propounded by Bloor whose inconsistency with cognitive science was evident from the accompanying assault on psychologism and the postulation of mental states. Its very blandness testifies to the falsification of the debate. The truism that there are social dimensions to science would hardly have generated the opposition and controversy evoked by the Strong Programme. How Bloor can now construe an attack on the reality of mental states as consistent with cognitive science remains a mystery he neglects to explain. Of course, Bloor's attempted rapprochement with cognitive science is also somewhat difficult to reconcile with his explicit endorsement of behaviourism.

Significantly, Bloor's sociological colleagues have reacted differently: their vehement attacks on cognitive science and artificial intelligence have been both telling and more ingenuous. Their strenuous attempts to discredit the claims of cognitive science have been to tacitly acknowledge the threat posed to the central sociological doctrines. Indeed, H.M. Collins (1991), among others, has been perfectly explicit on this point, seeing the claims of artificial intelligence (AI) as a crucial test case for the sociology of scientific knowledge.

CAUSES AND CASE STUDIES: 'BEYOND DISPUTE'

We have noted Bloor's contrast between 'teleological' accounts on the one hand with '*the* causal' model – by which Bloor means a sociological one, as if there could be no other kind of causal account. This is no inadvertent slip, for the issue concerning the causes of behaviour is at the heart of the Strong Programme – and the source of its most serious shortcomings. The foregoing issues are seen in a clearer light when it is understood that the specific interest of the sociological programme lies in the purported causal link between social milieu and theory contents. It is this link which constitutes the grounds for an historicist, contextualist, relativist view of scientific theories.

From its beginnings, the phenomena of central interest for the Edinburgh school have been 'the conditions which bring about belief or states of knowledge' (Bloor 1976, p. 4; 1983, p. 137) and especially the causation of belief or theory content. The extensive body of case studies repeatedly invoked by sociologists to answer their critics has been taken to establish the thesis that beliefs have *social* causes, in contradistinction to psychological ones. The equanimity, indeed seriousness, with which this thesis has been received suggests that there has been little appreciation of quite how scandalous the claim really is. I have attempted (Slezak 1989, 1991) to reveal the unnoticed enormity of the sociological programme which might, at best, be seen as a recidivist attempt to revive a discredited behaviourism, being propounded in apparent ignorance of the developments in relevant matters over the last thirty years. The claims for the social causation of belief constitute an utterly bizarre conception of human beings, and, in its recent sociological form, vastly more implausible than its earlier psychological counterpart.

Though seemingly free from *a priori* constraints, this causal tenet of the strong programme turns out to be, in fact, part of Bloor's peculiar view which assimilates the 'empirical' with the 'sociological' as if there could be no other kind of empirical study. For example, in reacting to the ideas of Lakatos, Bloor significantly sees the consequences of a 'rational reconstruction' as rendering science 'safe from the indignity of *empirical* explanation' (1976, p. 7, emphasis added), whereas for Lakatos only *sociological* explanations were intended to be excluded. Notice that Bloor is imputing to Lakatos the view that by offering a rational reconstruction, the beliefs are thereby shown to be lacking empirical explanation altogether! On this view ascribed to his philosophical opponents, Bloor complains further that '*empirical or sociological* explanations are confined to the irrational' (1976, p.7 emphasis added), – once again assimilating the two. This equating of empirical with sociological suggests either Bloor's failure to imagine that empirical explanation of belief might advert to other than sociological notions, or that it is Bloor, after all, who is making a stipulation to guarantee the truth of his doctrine.

The causal claim at the heart of the Strong Programme concerns such things as 'connections between the gross social structure of groups and the general form of the cosmologies to which they have subscribed' (Bloor 1976, p. 3). That is, the very cognitive content of the beliefs is claimed to be causally connected with immediate, local aspects of the social milieu. Of this general thesis, Bloor asserts 'The causal link is beyond dispute' (1976, p.3). Indeed, Bloor (1981) and Shapin (1979) were evidently unable to believe that anyone might question the causal claims of the Strong Programme except on the assumption that they must be unfamiliar with the extensive literature of the case studies. These were taken to provide the straightforward empirical evidence on which the Programme rests. However, the claims of social determination of beliefs are all the more extraordinary in view of the utter failure of these case studies to support them. The criticisms in question have challenged precisely the *bearing* of these studies on the causal claims, and so repeatedly citing the burgeoning literature is to entirely miss the point.[7]

Of course, scientific discoveries have always necessarily arisen in some social milieu or other, but this is merely a truism holding equally for much human activity not thought to have been actually caused in this way by social factors. The claims of the Strong Programme go beyond merely acknowledging a social context for scientific discoveries to asserting the bolder thesis that social factors are an irreducible component and causally efficacious determinants of the very content of theories. However, to the extent that social factors are indeed ubiquitous, it is not at all clear how one might distinguish mere co-occurrence from causal connection. Establishing a causal connection requires more than merely characterising in detail the social milieu which must have existed. These more stringent demands have not been met anywhere in the growing number of case studies emerging in the SSK literature. Nor has there been any awareness of a need to meet them by going beyond purely descriptive accounts of social contexts which happen to accompany some scientific work. Thus, although Steven Shapin has acknowledged that 'the task is the refinement and clarification of the *ways* in which scientific knowledge is to be referred to the various contextual factors and interests which produce it' (1979, p. 42), and that 'we need to ascertain the exact nature of the links between accounts of natural reality and the social order' (ibid.), nevertheless his case study of phrenology offers only a variety of anthropological approaches leading at best to a postulation of 'homologies' between society and theories which may serve as 'expressive symbolism' or perhaps function to further social interests in their 'context of use'. This falls far short of demonstrating the strong claims of social determination which abound in the rhetoric of programmatic statements and their 'social epistemology'. Indeed, the caveats and qualifications entered to the universality of even these weaker connections run the risk of rendering the claims empty, much less warranting Bloor's assertion that 'the causal link is beyond dispute' (1976, p. 3). Thus, it is a truism to assert, as Shapin does, merely that 'Culture [taken to include science] is developed and evaluated in particular historical situations' (1979, p. 65). The inescapable ubiquity of social contexts does not warrant the further claim that science cannot be understood apart from these historical contexts of use and social interests. Bunge has captured the essence of the sociological doctrine:

> Radical externalism is the thesis that all knowledge is social in content as well as in origin. In other words, tell me what kind of society you live in and I'll tell you what you think ... we should be able to read society off scientific theories ... (1991, p. 539-40).

It is this purported causal connection between the actual content of theories and specific, historical social contexts which is being challenged. Although offered as 'straightforward scientific hypotheses', we have noted that the claims of social determination of beliefs cannot be supported by the bald appeal to the case studies claimed to support them. As already noted, the stock reaction of the sociologists to their critics is simply to cite the case studies and to blame criticism on 'ignorance of this new literature' (Shapin 1979). Shapin undertakes to refute the accusations of empirical sterility by a lengthy recounting of the 'considerable empirical achievements' of the sociology of scientific knowledge (Shapin 1982, p.157-8). But he is simply begging the question with his advice that 'one can either debate the possibility of the sociology of scientific knowledge or one can do it' (1982). The point would be clear in the similar case of an astrologer who recommended getting on with the work of casting horoscopes rather than engaging in further 'debate'. Critics who question the inductive support of such horoscopes for the causal claims of astrology require a better answer, not more 'case studies'. Likewise, the mere ubiquity of social contexts for the occasion of any scientific belief is not ipso facto any better evidence of a causal link than the ubiquity of planetary configurations at any birth is evidence for astrological influence.

CAUSES AND COUNTERFACTUAL CONDITIONALS: NEWTON'S INVERSE CUBE LAW

What is the precise force and warrant of the generalisations cited in the SSK literature claiming a causal link between the contents of a theory and some feature of the social milieu? Although this question has been repeatedly posed, there is no evidence that it has ever been directly addressed, let alone answered, by proponents of the sociological approach. Thus, in a recent anthology, the editor Andrew Pickering yet again cites Shapin's (1979) phrenology paper as the 'classic study' and as 'exemplary work in SSK' which provides 'yet more documentation of the SSK thesis of social relativity' (1992, p. 6).

Genuine causal generalisations must support counterfactual conditionals, so that, had the cause not obtained, the effect would not have followed – or, given a different cause, a different effect would have ensued. Thus for example, we might ask 'Had the relevant social circumstances been other than they were, could Isaac Newton have propounded the inverse *cube* law of universal gravitation?' Merely to ask the question is to expose the profound difficulties, not to say absurdities, confronting the causal claims of SSK. In particular, we might ask again how the voluminous case studies have answered this kind of question.

The claims of social determination for theory contents lead to further awkward questions. The implausibility only becomes apparent when one looks beyond the coarse grain of the *recherché* 'homologies' to the detail of some work such as Newton's *Principia* or Chomsky's grammar and asks what social factors caused this or that particular theory. Following the model of Forman's (1971) much cited study which attributes the development of quantum physics to the milieu prevailing in Weimar Germany, we might inquire: Has Chomsky's latest linguistic theory of 'Government and Binding' arisen from influence of sado-masochism in politics? Likewise, we might ask: Did Gödel's 'Incompleteness' Theorem arise from some lacunae in the Viennese social order of 1930? These latter, admittedly facetious, examples parody the kinds of connections postulated by sociologists of science, but the distinction between parody and serious claims is difficult to discern.

If the specific contents of theories are supposed to have social causes, then embarrassing questions cannot be honestly avoided by recourse to vague generalities concerning 'homologies' with the *Zeitgeist*. The mismatch between social contexts and the detailed contents of scientific theories suggests the prima facie implausibility of the causal connection – this implausibility being itself only a symptom of a deeper problem concerning the utterly mysterious mechanism allegedly linking social factors with beliefs.

More generally, what of all the other scientific theories, with radically *different* contents which have emerged from the *same* social milieu? The question invites comparison with one put to astrology: namely, the analogous difficulty in explaining how people with similar planetary configurations at birth may have widely divergent personalities and fates. We might also ask: Do the specific social factors alleged to cause belief contents operate only on the discoverer, or do they also act to cause the acceptance of the same belief in all its followers (as Forman's study suggests)? Are we to suppose that my belief in Boyle's law has social causes similar to those which caused Boyle's belief? And then, what about my concurrent belief in the Heisenberg uncertainty principle?

Any attempt to confront such questions seriously reveals the obvious difficulties of mapping social factors onto the contents of beliefs or theories. Of course, the case studies repeatedly cited by proponents of SSK clearly acknowledge the obligation to provide such a mapping, for they actually purport to demonstrate just such connections. After all, such a mapping is the very point of their causal claim. However, these causal claims gain such plausibility as they might have through their very tenuousness and indirectness. The connections typically postulated are highly suggestive and even metaphorical: there is at best a kind of affinity claimed between the social context and the contents of the theory in question. Thus, Shapin cites 'homologies between society and nature' and an 'expressive symbolism' which can be exploited to serve social interests. So long as the contents of a theory are characterised only at the most superficial levels, such affinities are easy enough to discover or contrive. But the risk of vacuity in such claims is evident when Shapin extends the scope of homologies and expressive symbolism from what may be the *actual* experience of agents to their *ideals* and *aspirations*. There are now no constraints at all on how one might postulate the relevant social factors.

INTEREST THEORY AS PSYCHOLOGICAL

The shift to an 'interest' theory might be seen as a subtle refinement of the sociological view avoiding the problems with the paranormal character of its causal claims. Unquestionably, positing social interests as mediating between context and content avoids the difficulties with the 'astrological' character of social determinism, however this improvement is won at the cost of renouncing the original point and radicalism of the strong program. Citing social interests as causes is more plausible than social contexts per se, – but, of course, interests are nothing if not psychological! Interests are psychological entities akin to motives, goals, purposes and desires, – all constructs in good standing among contemporary cognitive psychologists.

Worse still for SSK, this move threatens not merely 'psychologism', but also triviality. The shift to interests removes the issue from the realm of the 'paranormal' but it nevertheless remains a matter of plausibility in competition with alternatives – now *other* psychological factors. As before, merely citing interests on the part of scientists, even those unquestionably motivating their work, is not yet to establish these interests as causes of theory contents. At best, the interests may be among the many other necessary conditions for the development of some theory, but to count as a cause they must also be *sufficient*. It is here that the interest approach is likely to founder, since there are many other (and rather more plausible) candidates for the sufficient psychological conditions of scientific discovery.

Though a social interest theory, admittedly a mentalist account, is merely implausible, it is also, I suggest, potentially trivial. This possibility emerges with a further attempt to salvage it. Vastly more plausible as causal 'interests' are those which promote, not so much personal, professional or social goals, but rather the interest in understanding, in explanatory success or perhaps in finding the 'truth'. These are what Shapin (1979) has called 'technical-instrumental' interests in 'prediction and control'. Formulated in this way, I believe 'interest theory' to be above reproach, but it is a Pyrrhic victory for the sociologist of knowledge. The theory has now lost all of its original radical content gained through the *contrast* with traditional views. To say that the scientist acts out of an interest in pursuing understanding or explanation for the purposes of 'prediction and control' is true but trivialises the proprietary sense of 'interest' the sociologist requires. Stating social interest theory in this way is to disguise the fact that the interests are no longer social – it simply poses the traditional question we should like to answer. That is, we would like to know precisely how scientific theories can satisfy these very interests. This is just the question which 'rationalist' philosophers have been traditionally asking and to which the work in cognitive science offers our best answers.

UNDERDETERMINATION AND THEORY 'CHOICE'

One consideration, above all, has been widely taken to warrant the appeal to sociological factors in the explanation of scientific theories. This is an argument which attempts to exploit the now familiar Quine-Duhem insight concerning the underdetermination of theory by evidence. Indeed, a non-sequitur involving this thesis has become one of the foundational tenets of the social constructivist enterprise. When distilled to its essence, the entire 'argument' underlying Bloor's (1976) *Knowledge and Social Imagery* is just this spurious inference from underdetermination to social construction.

The suggestion has been that the thesis of underdetermination of theories by their evidence justifies invoking sociological factors to explain a particular theory choice. The appeal of this argument is apparent: If the evidence is not itself sufficient to uniquely determine theory choice among the many inductively equivalent ones, then it is necessary to account for the 'choices' which are actually made by invoking some further considerations. A precisely analogous problem arises in the case of perception where the data of sensory input does not uniquely determine any interpretation. Thus, perception, like science, may be understood as the formation of hypotheses to account for the available data, and Collins (1991) has explicitly taken the perceptual problem as an illustration of social determination. In perception, as in science, there is no absolute fact of the matter regarding the 'correct' interpretation of any data. In the familiar textbook example, the black and white blobs might be a meaningless pattern, or they might be interpretable as a Dalmatian dog. Ingenuity might permit construing the pattern in yet other ways and there is no one which is correct in any sense. From this circumstance Collins infers that any interpretation must be socially determined, just as in the analogous case of scientific theories. However, the preference for seeing a Dalmatian must surely invoke psychological mechanisms, and the extent to which social factors play a part, if at all, must be an open empirical question rather than an *a priori* constraint on any possible solution.

Setting the pattern for subsequent discussions, Bloor (1976) utilizes this issue as one of the crucial arguments in his book. Following a rehearsal of familiar ideas concerning the theory-ladenness of perception, Bloor simply announces 'the social component in all this is clear and irreducible' (1976, p. 28). This typically bold assertion is supposed to follow from the arguments about theory ladenness, but, in fact, Bloor makes absolutely no connection between these ideas at all. Theory-ladenness is, of course, the other side of the underdetermination coin. Bloor appears to think that the phenomenon of theory-laden perception must *ipso facto* establish that the theories with which perceptions are laden must be sociological ones. Evidently no argument is thought necessary and Bloor gives none. In view of the foundational status which this book and this argument have acquired, the situation is sufficiently peculiar to deserve emphasis: The problem with Bloor's discussion is not merely that his arguments are weak or open to challenge in some way. Rather, Bloor actually gives no argument of any kind whatsoever.

However, notwithstanding the widespread resort to the Quinean thesis of underdetermination, it is entirely indifferent in its bearing on the sociology of scientific knowledge. Underdetermination as such is completely neutral among the possible alternative resources which must be invoked beyond conformity with the evidence to explain theory choice. For this reason, merely citing the Quinean thesis in support of the sociology of scientific knowledge is entirely gratuitous, and asserting the decisive role of social factors question begging. Clearly, it has to be shown *independently* why it must be *social* factors rather than some others which are, in fact, the operative ones in determining theory choice among the possible alternatives consistent with the evidence. Patently this has not been demonstrated merely by referring to underdetermination as such. Even if underdetermination does point to the need to invoke some kind of 'extra-empirical' factors to explain theory choice, this does not yet provide support for any particular candidate. The same argument could be offered with equal force in favour of astrological influences or mystical revelation as the decisive causes of belief. Nothing in Quine's thesis entails either that there must be any 'extra-empirical' factors, nor that these must be social. The failure of any set of *observations* on their own to uniquely determine an explanation says nothing about the many other considerations which will be brought to bear including, for example, simplicity, coherence, comprehensiveness, consistency with other theories etc.

It is worth observing that undoubtedly part of the problem here arises from an excessively literal construal of 'theory choice'. The formal, logical point of the underdetermination thesis, cannot be seriously read as entailing an actual choice among inductively equivalent alternative scientific theories. Historians above all should be aware of this as a fact about scientific discovery. The problem will be typically to find even a *single* theory which is consistent with the observations, however many there must be in principle. This obvious point appears to have been overlooked in the attempts to exploit the Quinean thesis, but it should be clear that what is termed theory 'choice' is more appropriately described as theory invention or discovery. It is only a misguided conception of the scientist as literally faced with a choice among observationally equivalent alternatives which lends any credence to the sociological thesis. The observational evidence does not exhaust the scope of rational, scientific, empirical, aesthetic or other internal considerations which might be brought to bear. Consistent with the underdetermination is the application of various problem solving strategies, heuristics and reasoning processes which constitute the psychology of scientific discovery – the subject of a considerable research enterprise today (Langley et al., 1987, Holland et al. 1986, Shrager et al. 1990). Compelling though the argument from underdetermination appears at first sight, its support for sociological determination of theories is entirely illusory.

DURKHEIM *DÉJÀ VU*

The ideas at the heart of the sociology of scientific knowledge are not new. Durkheim and Mauss (1903/1963) in their work *Primitive Classification* claimed that the cosmologies of groups reflect precise features of their social structure. For example, Durkheim and Mauss elaborate the case that 'what we find among the Zuni is a veritable arrangement of the universe' (1963, p. 43). According to Durkheim and Mauss, for the Zuni everything in the universe is assigned to seven spatial regions and 'this division of the world is exactly the same as that of the clans within the pueblo' (1963, p. 44). In his paper 'Revisiting Durkheim and Mauss', Bloor (1982) invokes them in support of 'one of the central propositions of the sociology of knowledge' (1982, p. 267) – namely, their view that 'the classification of things reproduces the classification of men'. Bloor recommends that Durkheim and Mauss should be rehabilitated after having been consigned to the history books, since their work is important for 'showing not merely how society influences knowledge, but how it is constitutive of it' (p. 297). However, if we take Bloor's invitation seriously and revisit *Primitive Classification* – including the introduction by Rodney Needham favourably mentioned by Bloor – we discover that the Strong Programme emulates *Primitive Classification* to the extent of exactly reproducing its severe shortcomings. Writing thirty years ago, Needham makes trenchant criticisms of Durkheim and Mauss which are identical with those which have been levelled against Bloor's Strong Programme. Remarkably, Bloor neglects to reveal the precise nature and severity of Needham's criticisms. Needham draws attention to Durkheim's 'tendency to argument by *petitio principii*' in his claim that the classification of non-social things 'reproduces' the classification of people – the claim which Bloor characterizes without demurral as a 'bold unifying principle' but which Needham describes as an unwarranted, abrupt inference and logical error which flaws the entire work. On the alleged parallelism between primitive societies and their concepts, Needham writes

> Now society is alleged to be the model on which classification is based, yet in society after society examined no formal correspondence can be shown to exist. Different forms of classification are found with identical types of social organization, and similar forms with different types of society... There is very little sign of the constant correspondence of symbolic classification with social order which the argument leads one to expect, and which indeed the argument is intended to explain. (1963, p. xvi)

In the same vein, Needham notes further that on one claim their 'evidences on this point lend their argument no support whatever' and on another claim 'nowhere in the course of their argument do the authors report the slightest empirical evidence, from any society of any form, which might justify their statement' (xxii). Needham's judgement is considerably more damning than Bloor reveals, suggesting 'Durkheim and Mauss's entire venture to have been misconceived' (xxvi). In view of the more recent airing of identical concerns, the following remarks are worth quoting in full:

> Yet all such particular objections of logic and method fade in significance before two criticisms which apply generally to the entire argument. One is that there is no logical necessity to postulate a causal connexion between society and symbolic classification, and in the absence of factual indications to this effect there are no grounds for attempting to do so... If we allow ourselves to be guided by the facts themselves, i.e. by the correspondences, we have to conclude that there are no empirical grounds for a causal explanation. In no single case is there any compulsion to believe that society is the cause or even the model of the classification; and it is only the strength of their preoccupation with cause that leads Durkheim and Mauss to cast their argument and present the facts as though this were the case (xxiv–xxv)

These remarks take on special significance in light of the fact that it has been precisely the causal claim of the Strong Programme which has been the foundational tenet repeatedly asserted and repeatedly challenged. Needham continues, noting that Durkheim and Mauss,' are explicitly concerned to propound a causal theory, and it is this which they equally evidently fail to establish' (xxv). As noted earlier, aside from the question of causation, recent criticisms of the Strong Programme (Slezak 1989) have emphasized the crucial mismatch between scientific theories and the societies they are supposed to mirror. Thus, it is striking to see Needham earlier draw attention to the extensive evidence which actually suggests a conclusion exactly the reverse of that which Durkheim and Mauss suppose. ' That is, forms of classification and modes of symbolic thought display very many more similarities than do the societies in which they are found' (xxvi). Needham' s sober judgement is:

> We have to conclude that Durkheim and Mauss's argument is logically fallacious, and that it is methodologically unsound. There are grave reasons, indeed, to deny it any validity whatever. (xix)

On one central, but typical, claim Needham expresses an attitude to the work of Durkheim and Mauss which is significantly different from Bloor's. Needham says 'It is difficult not to recoil in dismay' from their 'unevidenced and unreasoned' explanations for the complexities of social and symbolic classification (xxiii).

CONCLUSION: WHERE TO PUT A SOUFFLÉ

> Education has two purposes: on the one hand to form the mind, on the other hand to train the citizen. The Athenians concentrated on the former, the Spartans on the latter. The Spartans won, but the Athenians were remembered.
> Bertrand Russell 1931 p.251.

One cannot help being struck by the difference in tone when reading sociologists writing about science by comparison with the likes of Bertrand Russell (1931), Jacob Bronowski (1965) and, not least, Albert Einstein (1954). Where Einstein expresses a deep reverence and awe for the comprehensibility of the world, and Bronowski articulates the inspirational, aesthetic dimension shared between science and art, and where G.H. Hardy (1969) explains the transcendent beauty of mathematical ideas, the sociologists see something less exalted. Consistent with their position that science is not essentially different from other merely social practices, they seem blind to the deep intellectual qualities, the inspiring ideas, about which many have rhapsodized. Where Chomsky sees the human mind striving at the limits of its 'cognitive capacity' to attain the creative insights of the scientist or artist, Brannigan (1981) sees genius as a matter of definition and nothing more than historically contingent social attribution. On this account, presumably Isaac Newton was just lucky to be in the right place at the right time.

Of course, from the 'externalist' perspective one is systematically precluded from appreciating the intellectual dimensions of the enterprise. Correspondingly, one's conception of the value and goals of a science education will be different. On one view, a science education would aim to cultivate independent, critical thought and the pleasures of intellectual curiosity. An alternative view is expressed by Collins and Pinch (1993) in their recent book under the heading of 'Science Education':

> Finally we come to science education in schools. It is nice to know the content of science – it helps one to do a lot of things such as repair the car, wire a plug, build a model aeroplane, use a personal computer to some effect, know where in the oven to put a soufflé, lower one's energy bills, disinfect a wound, repair the kettle, avoid blowing oneself up with the gas cooker, and much much more. (1993, p.150)

No doubt. But one can't help feeling that something essential has been missed. Indeed, in this prosaic view, Collins and Pinch neglect the aspect of science which has been its most characteristic feature since its inception with the Milesian presocratics – the role of the creative mind in providing a deep understanding of the world. As if replying to Collins and Pinch, Guthrie (1962) explains the origins of modern science:

> ...the neighbours and in some things the teachers of the Greeks, were content when by trial and error they had a technique that worked. They proceeded to make use of it, and felt no interest in the further question why it worked... Here lies the fundamental difference between them and the Greeks. The Greeks asked 'Why?' and this interest in causes leads immediately to a further demand: the demand for generalization. (1962, p.35)

Thus, although it is undoubtedly 'nice to know' where to put a soufflé, it is the 'why' questions and the demand for explanatory theories which remain the crucial features of science. However, this creative explanatory dimension cannot be appreciated from a perspective which systematically rejects the role of reasons, evidence and cognitive factors as somehow irrelevant or inessential. Above all, it is difficult to imagine how a science education might neglect these factors even if only seeking to convey how to make a soufflé. Predictably, instead of encouraging the intellectual questions which drive a scientific curiosity, Collins and Pinch recommend that a science education attend to the social negotiation, 'myths' and 'tricks of professional frontier science' as 'the important thing' (p. 151).

It is a commonplace to say that we live in an age of science, but like most commonplaces this one is misleading if we consider the knowledge of the average citizen even in the most advanced and prosperous societies. It is not uncommon to find university students who are surprised to learn that heavy objects do not fall faster than lighter ones. Nor is it uncommon to find students who are unable to explain the scientific fallacies in the racial theories of Hitler's *Mein Kampf*. The latter example suggests the grounds for concern about the state of our science education. A generation ago C.P. Snow (1959) famously suggested that a scientifically literate person should be able to state the second law of thermodynamics but this is surely too optimistic nowadays as the evidence for a widespread scientific illiteracy reaches alarming proportions (see Matthews 1994); nearly half the population do not even know that the earth goes around the sun once a year.

It is undoubtedly 'nice to know' how to 'do a lot of things', but this conception of science does not address the fundamental problem of a science education which concerns *understanding*. The need for scientific understanding is not merely the affectation of an educated elite but of urgent practical importance for the wider community. For example, the state of public education is likely to determine how life on earth survives, if at all. In recent times a certain genre of commentary presumes to dismiss scientific alarms about environmental destruction as unfounded, hysterical ravings of greenie 'zealots'. In particular, the question of whether such claims are a hoax or 'pseudo-science' is a matter on which the general public deserves to have an educated opinion. Whether the warnings about the 'greenhouse effect', the destruction of the ozone layer and species extinctions are not to be heeded when they conflict with economic benefits is not inconsequential. It is in this sense that the purposes of education – to form the mind and to train the citizen – become jointly necessary.

Harvey Siegel (1989) has recently written 'a central aim of education . . . is the fostering of rationality, or its educational cognate, critical thinking' (1989, p. 21). The point is quite general. Herman and Chomsky (1988) have suggested that, unlike the case of totalitarian regimes, in democratic societies where people's actions are not controlled, it is necessary to control their ideas. The capacity for independent critical thought if widespread is often a threat to certain interests and, therefore, not encouraged by those who are in a position to manipulate public opinion through education or the mass media. Chomsky has documented the disturbing extent to which so-called 'intellectuals' through their failure to be critical and skeptical have become the agents for corporate and government interests rather than the common good. Instead of promoting and exemplifying the traditional intellectual values of disinterested critical inquiry, teachers and 'intellectuals' have become a key element in the 'manufacture of consent' in democratic societies through the creation of the 'necessary illusions' of which they are themselves the principal victims.

Faced with this, it is the responsibility of educators to provide what Chomsky has called 'intellectual self defence' against ideological mystification and obfuscation. It is in the light of this Socratic conception of education as providing tools for critical thought that the developments in social studies of science must be understood. Far from encouraging a capacity for clear thinking and critical inquiry, these developments depend upon a corruption of language and thought which Orwell (1984) described as 'like a cuttlefish squirting out ink' which serves as 'largely the defence of the indefensible' (1984, p.363). In his famous essay, Orwell was concerned with the 'special connection between politics and the debasement of language'. My charge has been that the enterprise of SSK and the wider post-modernist fashion for textualist, historicist relativism can be seen to corrupt the standards of critical thought and honest inquiry. The causes for concern are brought into stark relief when we notice the way in which essentially the same doctrines concerning the consensual nature of reality have been applied outside the domain of science. Clearly, the relativist idea that there can be no appeal beyond an historically given interpretive context has quite general application. Thus, we see essentially similar contextualist doctrines have led Baudrillard to deny the reality of the 1991 war in the Persian Gulf! He does so on the grounds that notions of truth, falsehood and reality are outmoded dogmas of enlightenment or positivist philosophies. Instead, we must recognize that there can be no access to any reality behind our images, texts, discourses and language games. Christopher Norris (1992) has provided a telling exposé of these extravagant, not to say obscene, ideas which are fundamentally the same as those of social constructivism – only more explicit about their implications. We can see the direct relevance of these ideas to SSK where Norris locates Baudrillard in the wider fashion for pragmatist, anti-foundationalist or consensus-based theories of knowledge, theories which take it pretty much for granted that 'truth' in any given situation can only be a matter of the values and beliefs that happen to prevail among members of some 'interpretive community'. (1992, p. 16)

Clearly, these are just the same ideas at the heart of the SSK enterprise in its various forms. It should not require argument to establish what Norris calls the 'intellectual and political bankruptcy', to say nothing of moral depravity, of doctrines which lead to such conclusions. Above all, there is a disturbing affinity between such views and those of revisionist historians who would deny the gas chambers of Auschwitz. This kind of historicist relativism serves to illustrate the consequences of the wholesale rejection of philosophical notions such as truth and reality. Baudrillard is only more explicit and consistent in this regard than the sociologists of science. The idea of Latour and Woolgar (to be explored in Part 2) that substances synthesized in a laboratory are socially negotiated constructions is identical with Baudrillard's textualist line according to which there is no reality behind the images, inscriptions and discourse concerning the gulf war. Just as Baudrillard sees history as a fictive construct, so Latour and Woolgar describe science as the social construction of fictions.

However, ultimately the intellectual and moral depravity of the sociology of science is of less interest that the broader consequences of these doctrines. Chomsky has chillingly noted that the level of culture that can be achieved in our own society is a life-and-death matter for large masses of suffering humanity. That is, the extent to which citizens are capable of exercising a capacity for informed, independent and critical thought has immense consequences for the lives of people elsewhere. It is by subverting this capacity that the seductive doctrines of SSK have an import going beyond their specific content.

In a frequently anthologized essay, Chomsky (1969) wrote:

> With respect to the responsibility of intellectuals there are ... disturbing questions. Intellectuals are in a position to expose the lies of governments, to analyze actions according to their causes and motives and often hidden intentions. In the Western world at least, they have the power that comes from political liberty, from access to information and freedom of expression. For a privileged minority, Western democracy provides the leisure, the facilities and the training to see the truth lying hidden behind the veil of distortion and misrepresentation, ideology and class interest through which the events of current history are presented to us. The responsibilities of intellectuals, then, are much deeper that ... the 'responsibility of peoples', given the unique privileges that intellectuals enjoy. (1969, p. 256)

This responsibility of intellectuals to speak the truth and to expose lies is also the responsibility of teachers to instil. Above all, it is the responsibility of science teaching to convey what Bronowski called 'the habit of truth' which is central to the scientific enterprise. Bronowski writes

> The values of science derive neither from the virtues of its members, nor from the finger-wagging codes of conduct by which every profession reminds itself to be good. They have

grown out of the practice of science, because they are the inescapable conditions for its practice. (1965, p. 60)

The values extolled and exemplified in the works of Bronowski, Russell, Einstein and the entire Western tradition inherited from the Milesians is only as robust as our capacity to impart it to the next generation. This is both the threat posed by current sociological doctrines and the challenge for science educators.

This journal has been founded with the important goal of promoting contributions to science education from the fields of history, philosophy and sociology of science (Matthews 1992a, p. 1). There could scarcely be a more fundamental contribution to science education than the one offered by constructivist sociological theories, since they purport to overturn the 'the very idea' of science as a distinctive intellectual enterprise with its special values. If the social constructivist views are right, then the goals of the journal are illusory, though this would surely be only one of the minor casualties of such revolutionary insights.

NOTES

1. The heading is a play on Popper's title '*The Poverty of Historicism*' (which was, in turn, an allusion to Marx's book the '*Poverty of Philosophy*' which was a play on Proudhon's *Philosophy of Poverty*.)
2. Arguably, just as the positivists' pastime of ridiculing abstruse philosophers such as Heidegger was ultimately indefensible, so Popper's dismissiveness may well be unfair to Hegel. Nonetheless, my point is not to justify the particular scholarly exegesis on which these attitudes are based, but to endorse the underlying intellectual and moral concerns. That is, whether or not Hegel's particular philosophy is entirely deserving of Popper's contempt is less important than the broader *concerns* which motivate Popper's writing. Even if a more balanced treatment of Hegel is warranted than Popper gives, the *ideas* that Popper attacks, may be guilty as charged - though they may not be Hegel's own. Again, if Popper's account is a caricature of Hegel, it is a caricature embraced by recent authors.
3. In Slezak(1993) I have documented the startling inadequacy of Bloor's response to Popper's argument. Where Popper (1966b) claims to have found a strictly formal, logical refutation peculiar to historicism, Bloor reports him as merely finding the kind of limitation shared by any empirical inquiry.
4. The work of Merton and others who had already formulated the ideas of the current sociology of science are largely ignored today, and so there is some irony in reading Merton's remarks written in 1949 which acknowledge the source of these ideas among his own antecedents. Merton writes 'The last generation has witnessed the emergence of a special field of sociological inquiry: the sociology of knowledge (*Wissenssoziologie*)' (1957, p.456). Noting the long history of the problems, Merton observes further:
    The antecedents of *Wissenssoziologie* only go to support Whitehead's observation that '... Everything of importance has been said before by somebody who did not discover it.' (1957, p. 456)
5. Bloor's territorial conception of the problem would not be worth remarking upon if it were not for his own subsequent accusation that philosophers' reactions to his thesis can only be explained by their own unworthy territorial motives. The final chapter of Bloor's second book boldly proclaims the new 'heirs to the subject that used to be called philos-

ophy'. He writes: 'My whole thesis could be summed up as the claim to have revealed the true identity of these heirs: they belong to the family of activities called the sociology of knowledge' (1983, p. 183). Bloor then observes 'I do not expect these conclusions to be received with much enthusiasm by philosophers' who might be expected to respond 'by drawing the boundaries of their discipline tightly around themselves'. As if there could be no sensible argument responding to his case, Bloor goes so far as to suggest that philosophers could only reply by 'stipulating' themselves correct 'because delicate questions of ownership and territoriality are involved' (1983, p. 184).

Of course, if Bloor's fundamental sociological thesis is correct, then the very accusation and contrast makes no sense, since there are, after all, no other grounds for disputing claims. In particular, there are no limitations on the application of political explanations which derive from 'the special nature of rationality, validity, truth or objectivity' (1976, p. 1). We see here the inevitable inconsistency, not to say hypocrisy, according to which the claims of rational argument are explicitly rejected while being appealed to tacitly. In this case, philosophers, being motivated by territorial concerns, must resort to stipulation because they have no compelling arguments. However, if Bloor's thesis is correct, then he should have no complaint on this score, there being no such thing as rational argument, evidence or logic. Aside from the absurdity of implying that there could be no sensible argument in response to Bloor's case, his pre-emptive attack on any possible philosophical replies only serves to reveal Bloor's own territorial ambitions. There can be little surprise if philosophers should find this style of writing irritating.

6. In the preface to the second edition of *Knowledge and Social Imagery* Bloor says 'attacks by critics have not convinced me of the need to give ground on any matter of substance' and he says 'I have resisted the temptation to alter the original presentation of the case for the sociology of knowledge, thought I have taken the opportunity to correct minor mistakes such as spelling errors. I have also made a few stylistic alterations...'. For further details see Slezak 1991, 1994a.

7. The foundational tenet concerning causal efficacy was directly raised in my own symposium article (Slezak 1989), and although Bloor explicitly cites this as an example of the 'emotive stereotypes' employed by critics, he neglects to address any of the issues raised. In the preface to the new edition of his book, Bloor still maintains that 'attacks by critics have not convinced me of the need to give ground on any matter of substance' (1991b, p. ix).

But a disinterested observer of the debate might like to hear how Bloor responds in detail to these particular charges which concern the very foundations of the Strong Programme. Instead, Bloor simply adverts once again to 'the empirical basis of the subject' as if this was unproblematic. Under these conditions, meaningful dialogue is simply impossible. Repeatedly intoning the very dogmas which are in dispute is empty substitute for rational debate. Ironically, it is Bloor who dismisses his critics as 'Those who content themselves with stereotypes, rather than attending to the precise details of what sociologists of knowledge have written' (1991b, p. 163). He charges these critics collectively with failing 'to grasp even the most central doctrines of the position they are attacking' (ibid).

REFERENCES

Ashmore, M.: 1993, 'The Theatre of the Blind: Starring a Promethean Prankster, a Phoney Phenomenon, a Prism, a Pocket and a Piece of Wood', *Social Studies of Science* 23, 67–106.

Barnes, B.: 1981, 'On the Hows and Whys of Cultural Change', *Social Studies of Science* 11, 481–98.

Bloor, D.: 1976, *Knowledge and Social Imagery*, Routledge & Kegan Paul, London.

Bloor, D.: 1981, 'The Strengths of the Strong Programme', *Philosophy of Social Sciences* **11**, 199–213.
Bloor, D.: 1982, 'Durkheim and Mauss Revisited: Classification and the Sociology of Knowledge', *Studies in History and Philosophy of Science* **13**(4), 267–297.
Bloor, D.: 1983, *Wittgenstein: A Social Theory of Knowledge*, Columbia University Press, New York.
Bloor, D.: 1991a, 'Ordinary Human Inference as Material for the Sociology of Knowledge', *Social Studies of Science* **21**, 129–39.
Bloor, D.: 1991b, *Knowledge and Social Imagery*, Second Edition, Chicago University Press, Chicago.
Brannigan, A.: 1981, *The Social Basis for Scientific Discoveries*, Cambridge University Press, Cambridge.
Bronowski, J.: 1965, *Science and Human Values*, Harper & Row, New York.
Brown, J. R.: 1989, *The Rational and the Social*, London, Routledge.
Bunge, M.: 1991, 'A Critical Examination of the New Sociology of Science, Part I', *Philosophy of the Social Sciences* **21**(4), 524–560.
Chomsky, N.: 1969, *American Power and the New Mandarins*, Penguin, Harmondsworth.
Collins, H. M. & Pinch, T.: 1992, *The Golem: What Everyone should Know about Science*, Cambridge University Press, Cambridge.
Collins, H. M.: 1991, *Artificial Experts: Social Knowledge and Intelligent Machines*, , MIT Press, Cambridge, MA.
Durkheim, E. & Mauss, M.: 1963, *Primitive Classification*, translated and edited with an introduction by R. Needham, The University of Chicago Press, Chicago.
Einstein, A.: 1954, *Ideas and Opinions*, Crown Publishers, New York.
Forman, P.: 1971, 'Weimar Culture, Causality and Quantum Theory, 1918–1927'. In R. McCormmach (ed.), *Historical Studies in the Physical Sciences*, Philadelphia, PA, University of Pennsylvania Press, pp.1–115.
Gieryn, T.F.: 1982, 'Relativist/Constructivist Programmes in the Sociology of Science: Redundance and Retreat' *Social Studies of Science* **12**, 279–97.
Guthrie, W.K.C.: 1962, *The History of Greek Philosophy: Vol. 1. The Earlier Presocratics and the Pythagoreans*, Cambridge, Cambridge University Press.
Hardy, G.H.: 1969, *A Mathematician's Apology*, Cambridge, Cambridge University Press.
Herman, E.S. & Chomsky, N.: 1988, *Manufacturing Consent*, New York, Pantheon Books.
Holland, J., Holyoak, K., Nisbett, R., & Thagard, P.: 1986, *Induction: Processes of Inference, Learning and Discovery*, Cambridge MA, MIT Press.
Langley P., Simon, H., Bradshaw, G., & Zytkow, J.: 1987, *Scientific Discovery: Computational Explorations of the Creative Processes*, Cambridge MA, MIT Press.
Latour, B. & Woolgar, S.: 1979, *Laboratory Life: The Social Construction of Scientific Facts*, London, Sage.
Latour, B. & Woolgar, S.: 1986, *Laboratory Life: The Construction of Scientific Facts*, 2nd Edition, Princeton, Princeton University Press.
Laudan, L.: 1977, *Progress and Its Problems: Towards a Theory of Scientific Growth*, Berkeley, University of California Press.
Laudan, L.: 1981, 'The Pseudo Science of Science', *Philosophy of the Social Sciences* **11**, 173–198. Reprinted in J.R. Brown (ed.), *Scientific Rationality: The Sociological Turn*, Dordrecht, Reidel, 1984).
Laudan, L.: 1990a, *Science and Relativism*, Chicago, The University of Chicago Press.
Laudan, L.: 1990b, 'Demystifying Underdetermination'. In C. Wade Savage (ed.), *Minnesota Studies in the Philosophy of Science*, *XIV*. Minneapolis, University of Minnesota Press.
MacIntyre, A.: 1988, 'Panel Discussion: Construction and Constraint'. In E. McMullin (ed.), *Construction and Constraint: The Shaping of Scientific Rationality*, Notre Dame, University of Notre Dame Press, pp.223–246.
Matthews, M.: 1992a, 'Editorial', *Science & Education* **1**(1), 1–9.
Matthews, M.: 1992b, 'History, Philosophy, and Science Teaching: The Present Rapprochement', *Science & Education* **1**(1), 11–47.

Matthews, M.: 1994, *Science Teaching: The Role of History and Philosophy of Science*, New York, Routledge
Merton, R.K.: 1942, 'Science and Technology in a Democratic Order', *Journal of Legal and Political Sociology* 1. Reprinted as 'Science and Democratic Social Structure' in his *Social Theory and Social Structure*, New York, Free Press, 1957.
Merton, R.K.: 1957, 'The Sociology of Knowledge'. In his *Social Theory and Social Structure*, New York, Free Press.
Mill, J. S.: 1961, 'On Liberty'. In *The Philosophy of John Stuart Mill*, edited by M. Cohen, New York, Random House, 185–319.
Nola, R.: 1991, 'Ordinary Human Inference as Refutation of the Strong Programme', *Social Studies of Science* 21, 107–29.
Norris, C.: 1992, *Uncritical Theory: Postmodernism, Intellectuals and the Gulf War*, London, Lawrence & Wishart.
Orwell, G.: 1984, *The Penguin Essays of George Orwell*, Harmondsworth, Penguin Books.
Pickering, A. (ed.): 1992, *Science as Practice and Culture*. Chicago, The University of Chicago Press.
Pinch, T.J.: 1993, 'Generations of SSK', *Social Studies of Science* 23, 363–73.
Pinch, T.J. & Collins, H.M.: 1984, 'Private Science and Public Knowledge: The Committee for the Scientific Investigation of the Paranormal and Its Use of the Literature', *Social Studies of Science* 14, 521–46.
Popper, K.R.: 1963, 'Back to the PreSocratics' In his *Conjectures and Refutations*, London, Routledge & Kegan Paul. pp.136–165.
Popper, K.R.: 1966a, *The Open Society and Its Enemies: Volume 2, Hegel and Marx*, London, Routledge & Kegan Paul.
Popper, K.R.: 1966b, *The Poverty of Historicism*, London: Routledge & Kegan Paul.
Russell, B.: 1931, *The Scientific Outlook*, London, George Allen & Unwin.
Russell, B.: 1946, *The History of Western Philosophy*, London, George Allen & Unwin.
Russell, B.: 1961, 'Education'. In R.E. Egner & L.E. Denonn (eds.), *The Basic Writings of Bertrand Russell*, New York, Simon & Schuster, pp.401–412.
Shapin. S.: 1979, 'Homo Phrenologicus: Anthropological Perspectives on an Historical Problem'. In B. Barnes & S. Shapin (eds.), *Natural Order: Historical Studies of Scientific Culture*, London, Sage Publications.
Shapin, S.: 1982, 'History of Science and Its Sociological Reconstructions', *History of Science* 20, 157–211.
Siegel, H.: 1989, 'The Rationality of Science, Critical Thinking and Science Education', *Synthese* 80(1), 9–42.
Shrager, J. & Langley, P, (eds.): 1990, *Computational Models of Scientific Discovery and Theory Formation*, San Mateo, Morgan Kaufmann.
Slezak, P.: 1989, 'Scientific Discovery by Computer as Refutation of the Strong Programme' *Social Studies of Science* 19(4), 563–600.
Slezak, P.: 1991, 'Bloor's Bluff: Behaviourism and the Strong Programme', *International Studies in the Philosophy of Science* 5(3), 241–256
Slezak, P.: 1994a, 'The Social Construction of Social Constructionism' *Inquiry*, forthcoming.
Slezak, P.: 1994b, 'A Second Look at Bloor's Knowledge and Social Imagery' *Philosophy of the Social Sciences*, forthcoming.
Snow, C.P.: 1969, *Two Cultures and A Second Look*, Cambridge, Cambridge University Press.
Stove, D.: 1991, *The Plato Cult and Other Philosophical Follies*. Oxford, Basil Blackwell.

Reflections on Peter Slezak and the 'Sociology of Scientific Knowledge'*

W. A SUCHTING†

ABSTRACT. The paper examines central parts of the first of two papers in this journal by Peter Slezak criticising 'sociology of scientific knowledge' and also considers, independently, some of the main philosophical issues raised by the sociologists of science, in particular David Bloor. The general conclusion is that each account alludes to different and crucial aspects of the nature of knowledge without, severally or jointly, being able to theorise them adequately. The appendix contains epistemological theses central to a more adequate theory of scientific knowledge.

> ... our Histories of six Thousand Moons make no Mention of any other, than the two great Empires of *Lilliput* and *Blefuscu*. Which mighty Powers have ... been engaged in a most obstinate War for six and thirty Moons past. It began upon the following Occasion. It is allowed on all Hands, that the primitive Way of breaking Eggs before we eat them, was upon the larger End: But ... the Emperor [of Lilliput] ... published an Edict, commanding all his Subjects, upon great Penalties, to break the smaller End of their Eggs. The People so resented this Law, that ... there have been six Rebellions raised on that Account ... These civil Commotions were constantly fomented by the Monarchs of *Blefuscu* ... It is computed, that eleven Thousand have, at several Times, suffered Death, rather than break Eggs at the smaller End. Many hundred large Volumes have published upon this Controversy ...
>
> Swift, *Gulliver's Travels*, Pt I, Ch. IV.

1. INTRODUCTION

Peter Slezak has recently published a brace of papers in this journal on 'Sociology of Scientific Knowledge and Science Education' (Slezak 1994, 1994a). They are principally concerned with the theme indicated by the first part of the title. He says that 'measured by external criteria, recent developments in the Sociology of Scientific Knowledge can boast considerable success' (Slezak 1994, p. 271), that its influence extends, both in principle and in fact, to the science classroom, and that this influence is pernicious (1994, almost *passim* from pp. 265f to 290f, 1994a, pp. 329, 331, 335, 344). The theme of 'science education' appears chiefly by reference to the latter. More specifically, the first paper is mainly about the 'Edinburgh School' of sociology of knowledge, and especially the work of David

*Abridged version of original paper.

Bloor, whose first book, *Knowledge and Social Imagery* (Bloor 1976), 'launched' it (1994, p. 271); the second is entirely about Latour and Woolgar's *Laboratory Life* (Latour & Woolgar 1986).

The present paper takes its point of departure from the first of Slezak's pieces (from now on unattributed page references will be to this), ignoring the second, because it seems to me to be essentially sound, as far as it goes. However, the discussion will ramify into an independent examination of some central issues of 'sociology of scientific knowledge' and specifically of David Bloor's version.[1] I shall, as far I can, avoid involvement in the minutiae of Slezak's controversy with Bloor (especially the exegetical minutiae),[2] my central aim being to discuss a number of fundamental theoretical issues which are raised by those controversies but go well beyond them.

The footnotes comprise references to the literature and also some points supplementary to ones raised in the main text, though the latter is meant to be self-contained and fully intelligible without reference to the former.

The paper is mainly critical and, unfortunately, because of the nature of the material pursued, has to be carried on much of the time by what Hume called 'the tedious lingring method'. But there are some constructive suggestions (especially in the appendix), which I hope to develop at another time and place.

2. SLEZAK'S PRESENTATION OF SSK

Slezak generally employs the term 'sociology of scientific knowledge', or 'SSK', as he himself abbreviates it (and as it will be referred to here from now on), to denote a particular direction or group of directions within the more general field of 'sociology of knowledge'. More specifically, 'SSK' is what he sometimes calls 'social constructivism' (e.g., pp. 265, 268, 269), or, to use Bloor's designation, 'the Strong Programme'. It is clear that Slezak concedes the legitimacy of certain endeavours in the field of 'sociology of knowledge' and it will help to reduce the likelihood of confusion concerning his conception of SSK if these endeavours are sorted out to start with.

Broadly speaking, the 'Strong Programme' in sociology of knowledge, or SSK, may be contrasted with a 'weak' programme. The latter includes at least two areas. I shall call them 'traditional' and 'moderate' sociology of knowledge.
1. 'Traditional' sociology of knowledge is based on the idea that, in the words of Robert Merton (1957, p. 459), cited by Slezak (p. 272), 'only error or illusion or unauthenticated belief ... [is] ... socially (historically) conditioned'. Or, in Slezak's own formulation, the idea is that

> the explanations for ideas may be sought in social factors only if they cannot be 'immanently' or rationally determined. That is, theories judged to be correct and

founded on rational considerations are not in need of sociological explanation in the way that false and irrational theories are ... manifestly absurd beliefs are taken to require some special accounting of a kind not sought in other cases. In this sense, traditional conceptions relegated sociology to the dross of science, to its residue of false and irrational beliefs (p. 274).[3]

I shall take this to mean, in brief, that according to 'traditional' sociology of knowledge social factors are invoked to explain the existence only of what is other than scientific beliefs.[4]

2. 'Moderate' sociology of knowledge is also characterised in the passage from Merton cited just above. He speaks here of 'the "Copernican revolution"' in sociology of knowledge, namely, 'the signal hypothesis that even truths were to be held socially accountable, were to be related to the historical society in which they emerged' (Merton 1957, p. 460).[5] This introduces another sort of application of sociology to science, which Slezak clearly regards as a legitimate enterprise,[6] namely, the study of what may be called compendiously the social conditions for the production – and presumably also what may be called, by analogy, the circulation and consumption – of scientific knowledge, as contrasted with the knowledge actually produced. Such conditions may include, for instance, the ways in which social developments pose problems for science to solve, the ways in which scientific institutions work *qua* institutions, and the ways in which scientific results are applied.[7]

This 'moderate' conception does not change in principle the theoretical framework of the traditional one, for both use a sharp distinction between what is and is not scientific. The difference is, briefly, that whilst the traditional view sees social factors only as obstructing science, the moderate view holds that it may assist scientific progress, though, to repeat, this does not touch science in itself, the 'content' of the science thus forwarded. So Slezak can regard this form of sociology of knowledge as a legitimate area of inquiry. However, the undisputed fact that science is always carried on within a social context 'does not warrant the further claim that science cannot be understood apart from these historical contexts of use and social interests' (p. 280). This portends the third – 'strong' – conception of sociology of knowledge, or SSK, which is what he rejects, unconditionally.

3. Slezak characterises SSK in a number of different ways, which may, however, for present purposes, be divided into two.
   (i) There are his own formulations which, I suggest, may be boiled down to the following. (a) The 'substantive', 'cognitive' 'contents of scientific theories are 'externally caused' by the social milieux in which they arise. (b) The former in turn 'reflect' the latter.[8] (c) The warrant for scientific beliefs is a matter of 'interests' and the struggle between them.[9] Further, I propose that the following is a reasonable gloss on these theses. (a) and (c) are theses about the epistemology of theories: (a) about the (causal) origins of what may be called the propositional content of scientific theories, and

(c) about the grounds upon which they are held. (b) is a thesis about the semantics, reference, 'ontology' of the content of theories.

(ii) He writes at one point (p. 273) thus:
Bloor states the four tenets of his sociology of knowledge as follows:

1. It would be causal, that is, concerned with the conditions which bring about belief or states of knowledge.
2. It would be impartial with respect to truth and falsity, rationality or irrationality, success or failure. Both sides of these dichotomies will require explanation.
3. It would be symmetrical in its style of explanation. The same types of cause would explain, say, true and false beliefs.
4. It would be reflexive. In principle its patterns of explanation would have to be applicable to sociology itself.

Slezak does not actually present this as a direct quotation from Bloor, but in fact it is an almost literal reproduction of a central passage (Bloor 1976, pp. 4f), where he says that these four points 'define what will be called the strong programme in the sociology of knowledge'.

3. SOME COMMENTS ON THE PRECEDING

If we compare Slezak's own characterisation of the conception he is concerned to criticise, which is essentially Bloor's 'strong programme', with Bloor's own presentation, as cited, the resemblance is not striking. As regards Slezak's (a), (b) and (c), there is nothing in Bloor's words about, respectively, 'external' causation, theories' 'reflecting' social causes, or 'interests'. This is not to say that the doctrines which result from attempts to realise this programme do not contain these elements, but only that they do not seem to be part of the programme itself. So it is perhaps best to regard Slezak's characterisation as involving both an interpretation of Bloor's programme and as taking what he (Slezak) considers as some of the main consequences of that programme as part of that programme itself. Whether such a telescoping is a wise or otherwise defensible procedure, it will be best, at least for the sake of clarity, to distinguish the two.

In the light of what has just been said I shall now propose my own formulation of Bloor's theses as set out above, in terms particularly relevant to the present context. In doing so, I shall omit tenet 4 from further consideration, because, though it is no doubt important, and has had a significant part in the controversies around this conception of SSK,[10] Slezak does not discuss it. Focussing then on tenets 1–3, and taking into account the general 'sociological' orientation of Bloor's program (not actually made explicit in his words as cited), I suggest that Bloor's conception may be said, at this stage anyway, to boil down to two main theses. (A good deal more will be said about the exact character of his views in later critical discussion.)

1. All aspects of belief or knowledge, and specifically truth/falsity, rationality/irrationality, success/failure, are subject to causal explanation, and indeed causal explanation of the same general sort.
2. Such causal explanation always contains a social component.

I shall henceforth refer to (1) as the 'Symmetry Thesis', or 'ST' for short, and to (2) as the 'Social Symmetry Thesis' or 'SST' for short. I shall assume, as being evident enough, that (1) does not entail (2), though (2) entails (1).

[NB. pp. 155-170 of 1997 original have been removed]

4. THE CHARGE OF RELATIVISM AGAINST SSK

Towards the end of his first paper Slezak speaks of 'the intellectual and moral depravity of the sociology of science' (p. 290), meaning by 'the sociology of science', more narrowly, SSK. This sums up with optimal brevity his charge against the latter. A little more specifically, the intellectual depravity alleged consists, fundamentally, in its 'historicist, contextualist, relativist view of scientific theories' (p. 279) – to put it briefly, its 'relativism' – and its moral depravity in its subversion of capacity for critical thought, which has, as its other side, encouragement given to indoctrination – in general, the rejection of the norms of conduct embodied in the 'ethos of science' summed up in Merton's well-known quartet of 'universalism, communism [in the sense, roughly, of community of endeavour, W.S.] distinterestedness and organized skepticism' (p. 270).

Now it is clear from Slezak's presentation that the charge of moral depravity rests, logically anyway, on the charge of intellectual depravity. In any case, that a certain theoretical position has certain moral consequences (desirable or undesirable) or encourages certain morally assessible attitudes or actions cannot, at least without some further argument, be considered an objection to that theoretical position *qua* theoretical. At any rate, I shall restrict myself in this paper to SSK in its purely theoretical aspect.

4.1. *The charge of epistemological relativism in general*

As has already been pointed out, the gravamen of Slezak's case against SSK in general, and against Bloor in particular, is that it involves epistemological relativism; indeed this is one of the central issues in the history of debate in the general area of 'sociology of knowledge', perhaps the crucial one.[11]

Slezak does not actually state what he means by 'relativism'. So I shall have to try to supply a relevant meaning. This is not at all a straightforward matter for a mere glance at the history of philosophy suffices to show that very many things have been meant by it. For present purposes I suggest that it mean the doctrine the first documented statement of which is by Protagoras: 'Man is the measure [*metron*] of all things, of the things that are that they are, and of the things that are not that they are not' (as cited in Diogenes Laertius, VIII.51). That is, reading the passage in the light of considerations of it by, amongst others, Plato (*Theaet.* 178b) and Aristotle (*Metaph.* 1053a31), that what is the case about the world depends on, and is hence relative to, what human beings say is the case (rather than conversely). So, even more briefly, the thesis of relativism is summed up in the title of Pirandello's well-known play entitled *Cosí è (se vi pare)*, or, as it has been well rendered, *Right You Are (If You Think You Are)*.

Furthermore, Slezak does not actually set out any explicit argument to the conclusion that Bloor's SSK involves 'relativism', its being seemingly assumed that, given the general character of SSK, it is more or less obvious that it does. But this is not so. For one thing, the rejection of the idea that there are invariant, necessary principles of rationality no more directly entails relativism than the fact that, for instance, there are no invariant, necessary principles of agriculture entails that all such principles are equally good. So it will be necessary to examine some of Bloor's actual arguments and this will in fact be the next main task.

Before getting down to this business it may be useful to make the point that even if it can be successfully demonstrated that a certain doctrine entails relativism this is by no means, in principle, the end of the matter. For the actual force of such a demonstration depends on whether relativism is judged to be characteristic of knowledge. It is only if it is not that the conclusion of the argument, insofar as the latter is acceptable, counts against the doctrine in question; if it is so taken, then the fact that the position does entail such relativism is a point in favour of it. However, there is a special difficulty here in trying to decide that issue, because it is precisely whether such relativism is characteristic of scientific knowledge that is one of the central issues in any reasonably comprehensive theory of knowledge. So there is a strong tendency for argumentation in this area to beg the question at issue. It follows that, broadly speaking, there are two possible strategies in this matter of epistemological relativism. One is simply to accept that it is indeed a consequence of the position in question and to regard this as a point in favour of the position in question. The other is to argue that it is not such a consequence.

4.2. *Bloor's response to charges of relativism*

Bloor's strategy seems designed to have it both ways. On the one hand, he concedes the charge of relativism. For example, he refers to a critic who is 'firmly against the symmetry requirement and its implied relativism' (p. 178), and if there is any ambiguity about this (as between what the critic takes the symmetry requirement to imply and what Bloor himself believes it to imply) it is removed when, a little later on, he describes as 'correct' a critic's charge that the strong programme is relativist (p. 180).

On the other hand, he seems to seek to undermine the seriousness of this concession.

In fact he in effect deals with the charge of relativism in dealing with a charge of idealism. In this latter regard Bloor reports that critics have argued as follows: 'Reference to the facts has to be denied in order to put true beliefs on a par with false beliefs, so they can be said to have the same kind of cause' (p. 173). I shall now follow his response to this charge step by step, and to this end I shall break up what he writes into successive passages (from pp. 173–175) which are not all thus separated in his own text.

1. To start with, Bloor asks: what are 'facts'? They can, he answers, be considered either (a) as 'what true statements ... are *about* ... . [o]bjects in the world', or (b) as 'what true statements *state*', the 'verbal formulations' of a), 'the verbal description that is given of' (a). In the case of (b), 'facts fall on the side of the "content" of propositional attitudes rather than on that of their "objects"'. In brief, 'facts' can be considered either as real states of affairs or as certain descriptions of them. I shall henceforth refer to these two senses as 'facts-as-objects' and 'facts-as-descriptions', respectively.

2. Of what kind are facts-as-descriptions? This question arises because facts-as-objects 'will in general impinge equally on those who have true and those who have false beliefs about them', so that whilst 'there is no question of disqualifying' facts-as-objects as 'possible causes' they 'do not ... suffice to explain' facts-as-descriptions, since, presumably, the same facts in the first sense can be given different descriptions and so be different facts in the second sense. So it is necessary to examine that subclass of 'the "content" of propositional attitudes' which consists of 'beliefs picked out by their truth and thus standing in a privileged relation to reality'. The question is: 'What is the class thus picked out?'.

3. One possibility is that it is 'a natural kind of belief, or something analogous to a natural kind', in which case philosophers would have 'discovered that there are two kinds of belief, distinguished by whether they possess or lack the property of corresponding to reality'. However, he rejects this on the ground that '[w]e can't play God and compare our understanding of reality with reality as it is in itself, and not as it is understood by us'.

It is now time to pause for a stocktaking of the moves made so far. I accept the distinction made in (1), and I do not contest the question posed in (2). What of (3)?

To start with, what is *meant* by denying that a true description/belief (henceforth simply 'true belief') is a 'natural kind'? Presumably this presupposes knowing what is meant by *affirming* that it is a 'natural kind'. This in turn presupposes what is meant by a 'kind' and then what marks out a specifically 'natural' kind. It may be said, fairly uncontroversially, that a 'kind' is, briefly, a class of items having a common character that differentiates members of this class from non-members. It is impossible to say anything both significant and uncontroversial about the idea of a 'natural' kind. I shall simply venture the hopefully fairly neutral suggestion, sanctioned by some traditional authority (e.g., J.S. Mill, *A System of Logic*, I.vii), that a specifically 'natural' kind is one whose members have, in addition to the defining property, an unlimited, or at least significant number of other properties in common. On this account, an instance of a chemical element, say, belongs to a 'natural' kind but one of a square thing does not.

Understanding matters thus, Bloor's denial that true beliefs do or can form a natural kind rests, in effect at least, on the claim that there is no way of discursively identifying the relevant common character, namely, a relationship of correspondence between such beliefs and facts-as-objects, and/or that items that do in fact share this common character lack the relevant other shared characters. In the present context the former claim is undoubtedly the operative one.

This claim seems to me to be correct. Indeed it may be defended even more strongly than Bloor does. For it can be argued not just that such a direct identification is unavailable, but that it is incoherent. This is because to test for the relationship we should have to know how reality is in itself, but to know this would be precisely not to know how it is in itself since it is then in a cognitive relation to us. (An appeal to 'intuition', or something 'immediate' of the sort cannot save the day, for 'knowledge', in any intelligible sense, presupposes some distinction between knower and known, otherwise the 'knower' does not *know* the 'known' but is *part of* the known.)

But this is by no means the end of the matter for someone who holds that true beliefs form a natural kind. For such a person will certainly make the common move of distinguishing between, on the one hand, the question of the 'nature' of true belief, what is 'constitutive' of it, what marks it out as a natural kind, namely, correspondence with the facts-as-objects, and, on the other, the epistemic question of evidential 'criteria' for the holding of that relationship. However, such criteria are ruled out by an argument similar to that just used. For establishing such criteria would still involve some direct knowledge of facts-as-objects: otherwise it would not be possible to know what the signs of true belief (facts-as-descriptions) are. So, though there is no *direct* move, of the sort Bloor makes, from a rejection of the idea of a direct ascertainment of the postulated relation of correspondence between the two sorts of 'fact' to a rejection of the idea that true beliefs form a natural kind, the supplementary move is easy to furnish.[12]

So far so good. We now come to the crucial step in Bloor's argument.

4. This is that the correct alternative to the idea that facts-as-descriptions (true beliefs) form a natural kind, is that they

> form a social kind ... a class like the class of valid banknotes, or the class of holders of

> the Victoria Cross, or the class of husbands. Their membership in this class is the result of how they are treated by other people ...[13]

This is amplified by the remark that 'the reason for that treatment will be practical, complicated, and itself part of reality'.

We must now consider this claim, namely, that there is a significant analogy between, on the one hand, the class of true beliefs and, on the other, the class of valid banknotes or of holders of the Victoria Cross or of husbands. I shall take the banknote example to stand for all three.

The following very simple argument would seem to show that there is no significant analogy between the two in the relevant respect. Prescinding from some qualifications that are irrelevant here, it is *both* a necessary *and* a sufficient condition for something to be a valid banknote simply that it be treated in certain ways by certain (groups of) people on the basis, ultimately, of a certain agreement (explicit or implied) between them; in brief, that it be the subject of a 'convention'. However, taking 'true' in a customary sense, it is *neither* a necessary *nor* a sufficient condition for a belief to be true that it be simply the subject of a convention. For we take it that there can or could be both a belief that is true though held by a minority of only one (indeed a description that is true but believed by nobody), and a belief that is false though universally entertained. It is true that there is a reason for people's treating certain pieces of paper or plastic and not others as valid banknotes and that the reason is indeed, as Bloor says, 'practical, complicated, and itself part of reality'. Poincaré's remark about geometry that it is 'conventional, yes, arbitrary, no' (1952, p. 110) applies here too. But this is basically irrelevant to the central issue, which is the alleged analogy with true belief.

Essentially the same point may be approached in the following way. On the one hand, membership in the class of valid banknotes can be changed, in principle anyway, *simply* by implementing a decision to change certain conventions – 'the Lord gave, and the Lord has taken away' – and by a change in behaviour following from this. But, on the other hand, membership in the class of true beliefs (or what are counted as true beliefs – I shall take this qualification for granted from now on) cannot be so changed.

The preceding line of argument seems so obvious and so decisive that it is difficult not to wonder whether it does not rest on some misunderstanding of the thesis against which it is directed, one the correction of which would render that argument ineffective. However, the only move that presents itself is that what is in question is simply a proposal for, or stipulation of an unorthodox necessary and sufficient condition for a belief to count as 'true', to count as 'knowledge', namely, that it be the subject of a certain convention. Nevertheless, the text being discussed gives no warrant for this reading. Moreover, its general slant is quite contrary to it: for instance, if this is what is in question why bother with the distinction between two senses of 'fact'?[14] More generally, such a move not only concedes the whole relativist case, but would involve such a fundamental revamping of the ordinary understanding of 'true belief' as, in effect, to subvert it altogether.

The suggested analogy between true beliefs and valid banknotes becomes even more deeply problematic in the passage that follows those cited and commented upon above.

But, Bloor asks on behalf of the reader, isn't this idealism after all? Surely this is a disguised way of saying that truth is all in the

> mind of the believer, or that it is just a projection of our collective attitudes?If this *is* a species of idealism it would be a form... that is compatible with an underlying materialism... at most, an idealism about the semantic dimension of current forms of realism, but not an attack on its ontological dimension. It would also be strictly limited in its scope. For notice: a banknote is ultimately a banknote because we collectively deem it to be so. For all that, it is a real thing with weight and substance and location. None of this materiality is denied by what has been said about its social status as a banknote. The same applies to the people who occupy a social role. They are flesh and blood. That material reality is not denied but presupposed by their social status.

Thus Bloor's response to the charge of idealism (and relativism is obviously a codefendant) is, in a nutshell, that banknotes are material things and so are the holders of beliefs, and therefore that a basic materialism is not being impugned. However, though this is true, it is completely beside the point.It is equally irrelevant that banknotes on the one hand, and the 'bearers', so to speak of beliefs – speech-sounds, ink-marks, states of computer discs, etc. – on the other hand, are all material. In fact, this remark permits the question at issue to be restated in a slightly different way. On the one hand, a valid banknote is (a) simply as a *banknote* just a material object, like a piece of paper or plastic, and (b) as a *valid* banknote, simply that material object as it functions within a context of certain conventions. On the other hand, a true belief is (a) simply as a *belief*, let us say a material item like a psychological state (of a central nervous system). The question is whether (b) as a *true* belief it is, to complete Bloor's analogy, simply that material item as it functions within a system of conventions. And that is the view against which the preceding argument has been directed.

Bloor concludes with what may be called a 'compatibility thesis':

> Where does this leave the charge that the sociological approach neglects the part played by the facts as causes of our belief about them? On the first meaning of this ambiguous accusation, where facts are objects, I have shown this is false. On the second meaning, where facts are the content of beliefs, the charge is, in a way, correct. Leaving aside certain subtleties, the content of a belief is not to be treated as the cause of the belief. But that is because it *is* the belief. Nevertheless, critics may feel... that they are getting contradictory signals from the sociologist about the causal role of facts. They are not. They are getting consistent answers to two quite distinct questions – one about the role of reality, the other about the status of reports of reality. They are just mistaking these answers for inconsistent responses to the same questions.

However, this final statement only muddies the water still further. The point at issue is not the role of facts-as-objects in the causation of facts-as-descriptions (fact-as-content-of-belief), but the role of the first with respect to the question of the character of that sub-class of the second that are *true*. It is, in a definite sense, correct that the content of a belief *is* the belief. But then the analogy between a valid banknote and a true belief becomes problematic, because it is not clear that a belief *qua* propositional content is material in anything like the sense in which a banknote is material. And if this is so then the further question of an analogy between the validity of a banknote and the truth of a belief cannot arise – *caedit quaestio*.

Altogether, critics are perfectly justified in believing that they have been short-changed, that Bloor's view is a form of relativism/idealism.

4.3. *Further discussion of Bloor's response to charges of relativism*

If the critical discussion so far of move (4) is cogent, then something went wrong in the movement from (1)–(3) to (4), and what is indicated is a search for some explanation that renders intelligible the many particular problems with Bloor's argument. Let us consider the matter step by step.

a) Bloor suggests – rightly, I think, as already indicated – that the question of knowledge, and in particular the question of truth, is, at least in the first instance, a matter of descriptions, that is, of concepts linked together into statements.

b) Concerning these concepts Bloor defends a Wittgensteinian account of meaning (see especially his 1983) he calls 'finitism',[15] and says ' is probably the most important single idea in the sociological vision of knowledge' (1991, p. 165).[16] Finitism consists, broadly speaking, of two parts. The first comprises the negative thesis that meaning does not determine use, because candidates for this use-determining function (e.g. mental images) can be shown not to be able to do the job. The second comprises the positive thesis that, on the contrary, use determines meaning, and, specifically, use within functionally specific 'language-games' which depend on the observance of socially determined and enforced conventions and which exercise the authority over usage that in the alternative, traditional view of meaning is (allegedly) exercised by inherent meanings, 'intensions', essences, and so on. This is so not only in cases where the convention in a sense creates its own object (e.g., valid banknotes, promising) but also in the case of, say, natural classifications, for instance, colours, where certain publically accessible standard cases ('paradigms') of the use of the colour-predicate in question have to be agreed upon. Uses which determine current meanings do not determine future uses and therefore meanings. To put the matter another way, meanings may be and are extended, but do not pre-determine an 'extension' in the sense of traditional (which here includes modern) logic. 'Given' meanings are extended through 'family resemblances', the interaction between 'criteria' and 'symptoms', and so on. This is, as was said to start with, a doctrine of *meaning* – more specifically, a 'social' theory of meaning.

c) Now it is important to underline the fact that, as the account has been developed so far, there is, in principle anyway, an *unlimited* number of ways in which a concept can be formed in the first place and in which it can be subsequently extended. But, of course, in fact, only one or at most a small number of these ways are actually in play at a certain time, for the sufficient reason that if this were not the case our cognitive systems would be chaotic, and that means would not in fact be cognitive systems at all. So there must be some constraints on the logically possible diversity that the account as developed so far allows for.

This situation is clearly set out in a paper in which Bloor makes use of Lakatos's account of mathematics in *Proofs and Refutations* in the context of his own theory of meaning. If, Bloor writes (1978, p. 251), on Lakatos's account

> the stream of potential counter-examples [*sc.* to a proposed concept-formation] is endless, then the processes whereby we accord, or fail to accord, recognition of them must also be endlessly at work. Without their remorseless operation and that of the forces which govern them, there would be neither order nor coherence in mathematical knowledge. Its classifications, its counterexamples and its theorems would have no agreed relations to one another... We can say that Lakatos has shown us that mathematics is something that has to be 'negotiated'. Logically it is totally underdetermined, but if it is to be real knowledge, something objective rather than a confusion of subjective opinion, then it must be determined somehow.

He goes on immediately:

> The answer is that it is socially determined in the course of negotiations: mathematics is whatever is the outcome, and nothing more (*loc. cit.*).[17]

But suppose, to borrow from Hume, 'we still carry on our sifting humour', and ask: what, if anything, determines the course and outcome of these 'negotiations'? That Bloor allows this question is shown by the fact that he proposes an answer to it, specifically one using an account by Mary Douglas founded upon the notions of 'grid' and 'group' (see, e.g., Bloor 1978, pp. 257ff, 1983, Ch. 7)[18].

I am not concerned here with the details of this account, and, more specifically, not with the details insofar as it is used for Bloor's purposes. What I am concerned with is the point that in making certain sorts of 'forms of life' the sole or even major determinant of the course of concept-formation in general Bloor is conflating a number of very different cases. More specifically, he is leaving out of account constraints on concept-formation that are 'objective' in a sense other than the social or intersubjectively valid. What this comes to, indeed, is that he is really offering not so much a social theory of *knowledge* as a social theory of *meaning*. This throws a new light on his view, expressed in a statement quoted a little earlier, that finitism is perhaps the principal element in his whole theory. I shall now try to spell these claims out a little more.

If we recur to the concept of a valid banknote as an example, then we can easily see that, at the most general level anyway, this is determined in just about a maximally conventional manner – second only, in this respect to, say, what counts in a certain country as the legally permissible side of the road on which to drive. In practice, of course, even the conventionality of the latter is not wholly unrestricted, that is, the choice is not entirely arbitrary. For instance, there are obvious practical reasons for having general rules and ones as simple as possible. This goes even more for valid banknotes. For instance, there are obvious practical constraints on their size, durability, value, susceptibility to being counterfeited, and so on. Naturally, all this is not to say that this constraints cannot change, for they can and do.

If we turn to other social arrangements, like laws governing marriage, it is equally easy to see that here again the element of mere conventionality is constrained by facts that render the actual arrangements far from purely arbitrary. Such facts include more or less deeply entrenched property-arrangements, moral attitudes and religious beliefs. Again, these obviously not only can but do change, though once more in a generally non-arbitrary way.

Despite the non-conventional elements in such cases, elements grounded in factual states of affairs, there is a definite sense in which there are no non-'negotiable' constraints on the relevant conventions, except at a very general level. For instance, no conceivable social arrangements can escape the constraints imposed by the fact that human beings are, at the very least, partly purely material items which can only survive and reproduce in certain sorts of material circumstances. In this regard it is important to keep in mind that whilst it is a matter of fact that certain conventions and not others are in place at a particular place and time, and that certain states of affairs and not others explain, wholly or partly these facts, the conventions themselves *qua* conventions are not facts.

If we take a leap now into the sphere of mathematics, the situation shows significant similarities and differences with those just considered. A concept-formation in pure mathematics is to some extent conventional, but this conventionality is also subject in practice, like the above cases, to certain constraints. Broadly speaking these are of two sorts that may be called intra- and inter-theoretic ones. The most general sorts of constraints of the first sort are those that operate to maximise both the unity and generality of a branch of mathematics, of a group of branches, or even of mathematics as a whole. Constraints of the second sort are exemplified by developments in pure mathematics occasioned by the need of an empirical theory or branch of the empirical sciences for an improved (more rigorous, more 'user-friendly', more general, etc.) mathematical formalism. It is true that the conferring of a high value on unified and general systems of mathematics and on empirical theories using mathematics is a feature of certain arrangements and not others, though it is also true that it is not itself conventional that only certain concept-formations and not others permit these values to be realised.

Let us turn finally to the empirical sciences. Here we can find elements very like that of valid banknotes in certain respects. For instance, at the most general level of consideration, what system of measurement (metric, imperial, etc.) is used is purely a matter for decision, though here again there are factual considerations which may influence that decision, for instance, relative ease of calculation. Again, there is a similarity in the fact that, just as, given a certain exchange-rate, there is a non-conventional ratio of convertibility of one currency into another, so, whilst what '1 metre/foot in length' signifies is specified by reference to certain relevant standards, conventionally agreed upon objects, the ratio of the measurements in the two systems of some other object is not (except in certain essentially trivial respects) a matter of convention. But this example also points to a lack of parallel between the two cases. For whilst the money-value of a given object may change without any change in the object, its length does not change (in a given reference-system) without some change in the object. This point to what may be called a higher – indeed absolute – order of factual constraint. Passing now to the question not simply of forms of measurement, but of empirical theories themselves, then it must be said that whilst it is true that at least in some cases (and this minimal assumption is all that is needed to make the point) alternative formalisms (sets of descriptions) may be used to represent a certain sort of situation, this freedom is always relative to consequences with regard to the form of representation chosen (simplicity, heuristic fertility, generality, susceptibility to be integrated with other, factually related, branches of science, and so on). At this point it is apposite to cite the earlier Wittgenstein against the follower of the later:

> ...the possibility of describing the world by means of Newtonian mechanics tells us nothing about the world: but what does tell us something about it is the precise *way* in which it is possible to describe it by these means... Mechanics is an attempt to construct according to a single plan all the *true* propositions that we need for the description of the world. The laws of physics, with all their logical apparatus, still speak, however indirectly, about the objects of the world (*Tractatus*, 6.342, 6.343, 6.3431).

All this gives us the very paradigm of what constitutes the idea of a 'fact of the matter' and of 'knowledge' in its central sense.

It is in relation to the matter of *change* in meanings (in 'language-games'), that it is perhaps most clearly evident that the phenomenon of knowledge cannot be seriously tackled without making reference to their function in mediating assertions about how the world in fact is – where 'fact' pertains to 'facts-as-objects'. This is not to say that all such changes are so explicable,or even that any are wholly thus explicable, but just that, in scientific contexts anyway, the ultimate determinant of acceptability is their adequacy to their factual subject-matters of the sentences that those meanings make possible. As Barnes – second to none in his commitment to the tenets of the 'Edinburgh School' – writes in a discussion of the Aristotelian concept of 'average speed' in relation to the Galilean concept of 'instantaneous velocity': 'There was no contradiction in the concept. People were persuaded to discard it, or to differentiate it, *by reference to the real world*' (Barnes 1982, p. 38 – his emphases).

5. A BALANCE SHEET

Knowledge is a result of a certain sort of causal interaction with the real world whereby the latter is represented within a framework of socially constituted cognitive forms.[19] Stated in the barest possible terms, controversies like that between Slezak and Bloor which has been the subject of the foregoing turn upon a failure to see the significance of each element in this formulation.

On the one hand, the Slezak position focuses, in effect, on inquiry's ultimate causal anchorage in the real ('facts-as-objects'), leaving out of sight the social determinants of knowledge as a discursive formation ('facts-as-descriptions'). His criticisms are basically made from a position that is largely just assumed, and hence essentially question-begging with respect to the positions criticised. But it is not possible at this stage of the game to rest arguments upon routine appeals to 'reason','evidence', 'intellectual appraisal', when these are among precisely the ideas that are being contested. There was a time when, against a background of broad consensus, and depending on what Russell once called 'a robust sense of reality', this procedure would have passed muster. However, this time has gone, at least for those who wish to contribute to forward-looking reflections on general questions relating to knowledge. To borrow a political metaphor, Slezak's position is conservative. To speak thus is not to condemn it unreservedly, for conservative positions often serve to remind us of things that should be conserved, neither forgotten nor undervalued, even if they are never sufficient as a response to new problems, and are, by themselves, ultimately obstructive to the discovery of adequate solutions.

On the other hand, the Bloor position focuses, in effect, on the discursive dimension of the process of production of knowledge ('facts-as-descriptions'), and specifically social determinants of that dimension, paying at best lip-service to the real ('facts-as-objects') as a determinant of knowledge. This leaves him in a thoroughly confused relativism (as indeed Slezak asserts but fails actually to show). In addition, his conception of the social determinants of knowledge is ultimately impoverished, jejune.[20]

In a broad sense, each position might be said to be right in what it affirms, wrong in what it denies, underplays or overlooks. Indeed, when in an ecumenical mood, one may be inclined to say that the two positions are complementary.[21] However, in fact neither is in possession of concepts adequate to grasping the sciences as historically specific social practices directed towards the cognitive appropriation of the real. Such an account remains to be constructed in detail. Nevertheless, the time for disputes about which is the uniquely privileged end at which to crack the problem of knowledge has now passed.

APPPENDIX

The following draws together a number of threads present in the preceding discussions and weaves them into a more constructively oriented pattern. Though space-constraints restrict the presentation to little more than a series of theses, what is said may suffice to indicate the general character of an approach of the sort desiderated at the end of section 10 above.[22]

1. Knowledge (in the sense of knowledge 'that' rather than knowledge 'how') is, in the first instance, actually or potentially, a discursive or semiotic formation of signs (in a very broad sense: words in ordinary and constructed languages, mathematical symbols, diagrams, etc.).[23]
2. Obviously this is only a necessary, not a sufficient condition (for instance, it does not distinguish a claim to knowledge from a work of fiction). In addition, there must be a claim to represent, however partially, and with whatever degree of directness, the character of a subject-matter distinct and separable from it or, for short, a 'real object'. The latter may be wholly or only partially distinct and separable from various forms of awareness – consider, for example, the distinction between a piece of paper as such, and the same piece of paper *qua* valid banknote. However, the second sort of object always ultimately depends for its existence on the first sort. The two might thus be called 'primary' and 'secondary' real objects. In the following remarks I shall, for the sake of simplicity, focus on the first sort.
3. However, the conjunction of these two necessary conditions does not constitute a sufficient condition for knowledge, because an item may satisfy both without constituting knowledge – for instance, a randomly selected option in a multiple choice question. So a claim to knowledge must have attached to it what may be called a 'warrant'.[24] These three severally necessary conditions may, for present purposes anyway, be regarded as jointly sufficient for knowledge.
4. Knowledge is an outcome of a causal process. With regard to the latter we can distinguish here, as in all such processes, (a) what is causally transformed so as to produce the knowledge-effect, and (b) what brought about that transformation. I shall call these, respectively, the 'raw material' and the 'means' of knowledge.[25]
5. Consider first the case of what may be called 'simple perceptual knowledge' (e.g., 'This is a table' or even 'This is an oblong, brown table'). Here the raw material is (roughly) a real object (table) and the means (again roughly) the human sensory apparatus. It is essential to note here that the means are systematically ambiguous in character, or

fall simultaneously under two different but non-exclusive sorts of descriptions. Firstly, they are parts of purely natural processes. However, secondly, insofar as we have to do with a cognitive process, they involve a system of signs, a language. So, at the simplest level, knowledge has two distinct but inseparable dimensions, namely, causal in the ordinary sense, and discursive. The sensory apparatus in the broad sense is the mediator between the context of the real object and the context of knowledge.[26]

6. Simple perceptual knowledge typically does not consist of isolated claims of the sort instanced above but of more or less integrated systems of judgements, related by inferences, some of which are conscious, some unconscious ('natural', 'spontaneous'), the whole forming what may be called 'everyday' knowledge. In this way judgements once made may serve as new 'raw material' for knowledge, in addition to real objects.

7. The warrant for claims to simple perceptual knowledge and, more broadly, everyday knowledge, is, in general, what may be called its 'reliability' as a means by which human beings orient themselves in the world (sustain themselves and reproduce).[27] This reliability rests on ordinary perception in roughly standarised conditions and on coherence between different knowledge-claims.

8. Everyday knowledge has a social dimension in numerous different respects. The most important is the discursive, semiotic domain itself where practical social activity secures reference in general, at least for the fundamental parts of the language,[28] and largely determines the particular systems of meanings (classifications, etc.). As well, the social order is responsible for producing new real objects of knowledge and new means of knowledge both sensory and material-instrumental.[29]

9. Though adequate to many of the tasks for which it is produced, everyday knowledge is limited and inadequate over wide ranges, even in its own terms, and certainly when measured by high standards of unity, generality and precision. A crucial moment here is the fact that its cognitive means are largely limited to a more or less unassisted sensory apparatus which restricts it to mainly qualitative knowledge of a scope which is restricted both extensively and intensively. The requirements of everyday life itself, not to speak of other factors, press towards ways of producing knowledge more adequate in the respects noted. The result is 'scientific' knowledge.

10. The centrally determining factor here is the development of new *means* for producing knowledge, the production occurring in an increasingly purposive way. These may be divided into material means on the one hand, and discursive ones on the other. In making this distinction it must not only be kept in mind that these are, as before, inseparable, though distinct, but that, as the sciences develop, the two moments inosculate more and more: new material means are in

general the vehicles of more and more advanced discursive formations, and the latter point the way to the former. (a) The new material means are, centrally, forms of instrumentation. These more and more free knowledge from the already mentioned limits of the ordinary sensory apparatus, *inter alia* permitting the acquisition of exact quantitative knowledge and, in general, relational knowledge. They also not only open up raw materials for knowledge hitherto closed, but themselves produce new raw materials (e.g., new chemical elements). (b) The new discursive means are characterised, in the broadest way, by their being expressed in a specialised, 'custom-built' way, increasingly free of the unclarities and restrictive practical orientations of everyday language, permitting levels of unity, generality and precision inaccessible to everyday knowledge. More specifically, they they may be divided, at a first approximation into two sorts, that may be called the 'substantive' and the 'methodological'. (i) The first includes, above all, new concepts (e.g., 'inertial motion', 'chemical element', 'species' – in the Darwinian sense – 'mode of production') which form a thread uniting successive theories in a certain domain. (ii) The second includes new methods of theory-formation (in generality ranging from special mathematical techniques to what what may be called 'ideals of explanation'[30]) as well as of inference and appraisal of empirical warrant. The distinction is, as just suggested, only a rough one, useful for gaining an initial orientation in the field, for at least two reasons. One is that, even considered synchronically, the elements show a complex overlapping of features, both between the sides of the division and also within each side; another is that, considering the matter diachronically, an element may change its place in the classification, so that acquired substantive results become parts of methodologies for acquiring further results, the whole forming a cognitive 'loop'.[31]

11. At every point, social determinants enter the process by which scientific knowledge is generated. They cannot even be catalogued here, but at best exemplified. (a) They are crucial, directly or indirectly, in what has already been called the 'conditions' for the production of knowledge in respect of such matters as the setting of cognitive goals (which may, according to context, both forward and restrict research), in the organisation of the cooperative production of knowledge,[32] and in its circulation and distribution. (b) They enter into the production of knowledge itself. One such point of entry is the technical-instrumental means of research. Another is the provision of models for the understanding of scientific subject-matters,[33] models which are, of course, subject to the best historically evolved principles of appraisal available, which principles themselves may have had a social origin.[34] Yet another is as the source of those conceptions of the ultimate 'point' of what they are doing which are typically of a social provenance.[35] The whole forms what may be called a 'mode of production' of knowledge.[36]

12. It is within the above framework that questions about 'rationality' – or, better, in order to diminish the risk of reification, questions about what is 'rational' – can be posed. Such questions primarily concern actions and should, in the context of knowledge-claims, be put in terms of the efficacy of certain means (including both statements and procedures) for achieving the end of producing knowledge.[37] In respect of cognitive means it is very tempting to speculate that there are ways of dealing with situations that are 'hard-wired' into human beings, presumably as internalisations of evolutionary developments. At the other extreme are means that are 'state of the art', but likely to be superseded with the growth of knowledge (including instrumentation and the like). Between these two are a diverse set of means (explanatory schemata, particular theories, and so on) which have developed historically and have 'proved their mettle' to various degrees. The cognitive means most highly entrenched at a time tend to be looked upon as rational *per se*, claims to rationality being found less strong the shorter the time-period concerned.[38] The temptation to which rationalisms of various stripes fall is to take (possible) 'hard-wired' means and those that have proved themselves most permanent as being of more than merely natural-social provenance and thus unchangeable. Both generally have a broadly ideological significance, that is, in the present case, the role of anchoring the natural in the non-natural or making what is in principle changeable into what is necessary, each move being in the service of some values, interests.[39]

NOTES

1. I think Slezak erred in putting all those discussed in his two papers over the common denominator of 'sociology of scientific knowledge'. Barnes, Bloor and Shapin, for instance, are in a quite different class from the others, and especially from Latour and Woolgar. In another paper (1994c, p. 142), Slezak in effect touches on this point: 'Where Latour and Woolgar evade criticism by adopting deconstructionist double-talk and affecting a posture of nihilistic indifference to the cogency of their own thesis, Bloor professes to adhere to the usual principles of scientific inquiry'.
2. 'As far as possible': but, for instance, I cannot entirely avoid touching here and there, even if, sometimes, only glancingly, on Slezak's claims, made here and there in the paper to be discussed, but in more detail in other places (most recently in 1994b and 1994c), that Bloor has (tacitly) changed his position in various significant ways, notably as between the first and second editions of *Knowledge and Social Imagery*. At any rate, I think that Bloor is essentially right in saying that the topics covered in the 'Afterword' to the second edition of the latter 'represent the main areas of dispute in the field' (1991, p. ix), and I shall take what he says on his own behalf here as the main canonical statement of his views.
3. See also the following statements, taken from three consecutive pages of the paper and arranged simply in the order of their occurrence there. '... scientific theories ... [are] founded on logic, reason and evidence ...' (p. 265). 'What is usually called the "plausibility" or warrant for beliefs is ... an intellectual or cognitive question ...'; '... [scientific] beliefs are intrinsically the products of ... "internal" considerations of evidence

and reason...' (p. 266). Beliefs are 'a matter of reasons, evidence and other rational considerations...', ' the ... consequence of... internal, cognitive, intellectual [factors]', ' acquired and justified through mental reasoning processes', 'the products of ... - intellectual appraisal...' (p. 267).
4. Examples are not supplied, but I assume that an early instance of what is in question might be Francis Bacon's theory of 'idols' (*Nov. Org.*, I, 38–68). – Actually, the passages, both in Merton and in Slezak, on which this brief formulation is based (including the one from Slezak just cited in the main text) contain many at least potentially confusing (if not confused) statements of position. One is a tendency to slide from talking about simply *false* statements to talking about 'irrational' ones, and another to identify – verbally anyway – false statements which are nevertheless scientific with pseudo-scientific ones. As regards the first, 'false' and 'irrational' belong to different *categories* (the one pertaining to truth-value, the other to evidence) and are logically indifferent to one another. As regards the second, there can be statements which count as 'scientific' though they are false, and, conversely, ones that are true but 'unscientific'. However, there is no space here to analyse in detail the passages in question further. The brief formulation in the main text to which this is a note is simply what I take Merton and Slezak to be meaning to say, whether or not they always succeed in saying it coherently and clearly.
5. This brief passage illustrates one of the confusions referred to in the previous note. For to say that on this view 'even truths were to be held socially accountable' suggests that on the 'traditional' view falsehoods *as such* 'were to be held socially accountable', whereas surely what was held to be so accountable was not these but what may be called 'arational' beliefs.
6. Slezak (p. 273) sees the Merton formulation just quoted as one of or at least as foreshadowing the next conception to be considered, namely, SSK. But this is incorrect, as may be seen by considering the approach actually exemplified in Merton's immensely influential 1970 [1938]. Confirmation of this is offered by the view of the matter in Bunge (1991) – which, together with Bunge (1992), Slezak himself praises as a 'masterful survey of the field' (1994b, p. 336) – where he quotes (nor quite accurately but sufficiently for present purposes) the passage from Merton cited in the text to which this is a note with the remark that this formulates 'the central general hypothesis' of the book in question (p. 531) and later on explicitly distinguishes this view from the 'post-Mertonian directions' represented by, among others, the 'strong program' (pp. 533f).
7. Once more, in order for the reader to have easy access to the protocols of interpretation, I shall list, simply in order of occurrence in the text, what seem to be the major relevant passages. 'Previously, sociological studies paid attention only to such things as institutional politics, citation patterns and other such peripheral social phenomena surrounding the production of science...' (p. 273). '... sociological explanations are sought for the ... circumstances of ... [the] production [of theories]...' (p. 274). '... the sociological thesis ... [that] ... there are social aspects to science ... that there are social dimensions to science ... scientific discoveries have always necessarily arisen in some social milieu or other ... [there is] a social context for scientific discoveries ... social factors are indeed ubiquitous ... the inescapable ubiquity of social contexts...' (pp. 279, 280). – For putative examples see, for instance, Cohen (1994, pp. 216ff, 308ff).
8. That a) and b) are logically independent theses can be seen easily enough. Thus there could be significant analogies (some degree of similarity) between, on the one hand, the structure of a social system, or some aspects of it, and, on the other, the character of the physical world in general, or aspects of it, and the former may be used as a model for understanding the latter. But this does not entail that this theory is 'about' the social structure rather than nature. (An obvious example is the structure of a market society or a machine like a clock, on the one hand, and the 'mechanistic' conception of nature. Cf. Gideon Freudenthal 1986, 1988.) Or, to take an analogy, perceptual judgements are partly caused – directly – by events in the central nervous system in turn ultimately caused by states of the physical world, but this does not mean that the judgements in

question are 'about' the psychological events rather than the physical world. (Such a false inference has been, of course, one of the origins of psychologistic forms of empiricism.)
9. The following are passages on which this brief summary is based, arranged simply in order of occurrence in Slezak's text. '... the claims of SSK for the external causation of scientific belief...' (p. 265). 'The doctrines...[that]... take scientific theories to reflect the social milieu in which they emerge and... beliefs are taken to be causal effects of the prevailing context' (p. 265). '[Doctrines according to which] [w]hat is usually called the "plausibility" or warrant for beliefs is... only a matter of the "balance of forces" and political allegiences' (p. 266). '[The idea that] beliefs are intrinsically the products of "external" factors such as social causes and interests...' (p. 266). 'The very notion of merit, as indeed the honorific title of "Science" itself, is repudiated... as reflecting only interests and power relations among different groups' (p. 266). '... the externalist conception of theories as caused by features of the social milieu... On this conception, beliefs are the causal consequences of external environmental, social factors... the actual contents or "ideas" of science are [on this view] merely an epiphenomenal by-product of... social processes... beliefs are the products of social contexts...' (p. 267). '... the radical idea... of considering the actual substantive content, the ideas, of scientific theories... the cognitive contents of theories... as an appropriate domain for sociological investigation' (p. 273). '... the specific interest... [of this program]... lies in the purported causal link between social milieu and theory contents' (p. 278). '... the very cognitive content of the beliefs is claimed to be causally connected with immediate, local aspects of the social milieu' (p. 279).
10. See, e.g., Hesse (1980, pp. 31, 42–45), Jennings (1984), and literature cited there.
11. For a placing of more recent debates within the 'sociology of knowledge' tradition see Stehr & Meja (1982).
12. The key point in this argument is part of the burden of those superlative first few paragraphs of the Introduction to Hegel's *Phenomenology of Spirit*. Of course, Hegel eventually arrives at the essentially unintelligible idea of the identical subject-object. – In general Slezak tends to underestimate the significance of the idea that 'all we have, and all that we need, are our theories and our experience of the world' (Bloor cited in Slezak 1994b, p. 347) which is, he writes here, 'an uncontroversial view... a commonplace, shared on all sides'. This claim is simply absurd, as is that on the preceding page, regarding what is, in effect, the correspondence theory of truth, that 'it is unclear [*sc.* from Bloor's texts] who is alleged to hold this view'. I venture to name one name straightaway: none other than Slezak's much admired Popper. But it is in fact virtually an article of faith among *bien-pensant* 'realists'.
13. With this passage cf. esp. Bloor (1996), where the Wittgensteinian connection as regards the analogy of a true belief and a valid banknote is clear. This comparison goes back at least as far as William James (1955 [1907], p. 137): 'Truth lives... for the most part on a credit system. Our thoughts and beliefs "pass" so long as nothing challenges them, just as bank-notes pass so long as nobody refuses them...' Shapin quotes part of this passage, with approbation, in his (1994), p. 6.
14. However, it is true that in Barnes & Bloor (1982), there is the following passage (p. 22, n. 5): 'We refer to any collectively accepted system of belief as "knowledge"'. (See also, most recently, another adherent of the 'Edinburgh School', Shapin (1994), p. 4: 'For historians, cultural anthropologists, and sociologists of knowledge, the treatment of truth as accepted belief counts as a maxim of method, and rightly so'.)
15. The term seems to derive from Hesse (1974), esp. Chs. 8, 12. See also Barnes (e.g., 1982, esp. pp. 27ff).
16. Barnes (1982, p. 39) affirms: 'Finitism implies a thoroughgoing sociological treatment of knowledge and cognition'.
17. But it is relevant to note here that Bloor writes of *modus ponens* that 'the widespread tendency to argue in this form is because the pattern is innate... is a feature of our natural rationality.' The sociological aspect enters at the point of codifying and enforcing

this pattern through its erection into a 'cognitive institution'. 'As a logical convention it will now be subject to special protection, e.g., from counterexamples and anomalies in its application' (1991, p. 182). Thus the 'aura of the absolute that surrounds such principles as *modus ponens*' must have come from the social contrivances that constituted their special status. When we feel their compelling and obligatory character it is cultural tradition and convention to which we are responding. The '"realm of necessity", therefore, turns out to be the social realm' (1991, p. 183). – Another profound irony about the work of Lakatos, in addition to that pointed out in n. 19 above, is the fact that, as Bloor makes very clear in his use of it in his (1978), Lakatos's (1976) is a very paradigm of the employment of an approach to meaning according to which, roughly, 'meaning is constructed as we go along' (Bloor 1991, p. 164), an approach due, proximately at least, to Wittgenstein (Bloor 1983). ('Proximately': Bloor 1976, pp. 118ff points to also J.S. Mill with his doctrine that inference proceeds from particulars to particulars [*A System of Logic*, III.iii.3].) However, Lakatos had nothing but hatred and contempt for Wittgenstein and those influenced by him, like Toulmin. See Lakatos (1978, Vol. 2, Ch. 11, esp. pp. 224–228, 235f, 240f).

18. This may be seen as Bloor's attempt to fill a gap in Wittgenstein's thinking. As he points out (1983, p. 48), Wittgenstein simply says that language-games just change 'spontaneously' (*Remarks on the Foundations of Mathematics* [1956], III, 23); but, of course, this just amounts to rejecting the question. Bloor himself also gestures in the direction of 'needs', 'interests' (1983, pp. 46–49, 133–136), a direction criticised previously in this paper. The direction mentioned in the text seems more promising (even if, I believe, quite inadequate) and hence I have concentrated upon it here.

19. See the appendix for a little more detail on this.

20. Thus Bloor writes of 'the interaction that would create a society', and of 'negotiation' as 'a social process' (1991, pp. 168, 169), and defines the province of sociology as '[t]he science which studies the conventions which always attend and structure the expression of our urges and capacities' (1988, p. 67). It is astonishing that a leading representative of 'sociology of scientific knowledge' should have no more than this vague conception of the social. Already in (1978), p. 272, n. 60, Bloor acknowledged the 'lack of a unifying theory'; the other quotations in this note show that the situation has not changed since.

21. The two positions are in fact complementary in the sense of being variants of a single underlying 'problematic', however paradoxical this might appear in the light of Slezak's emphasis on 'reason' and Bloor's equally aggressive 'naturalism', which in his case is given a positivistic 'reading'. For central to positivism is an attitude which places in the foreground what is actually the case, one that, correlatively, pushes into the background the idea of explanation and, in connection with this, rejects or at least radically 'reparses' the idea of cause. Of course, for positivism what is actually the case is what is observable, in one or other of the many senses of that term. But rationalism also places in the foreground what is the case, though here it is 'reason', 'rationality' and in every significant sense exempts it from causal explanation. (In this regard I agree completely with Bloor's point that philosophers *typically* try to erect barriers to causal explanation – e.g., 1983, pp. 74, 76f. This points to why it is of the greatest importance to distinguish, as I did to start with, between Bloor's basic program and the way he carries it through, for the two should be evaluated differently.) In this way, traditional positivism and rationalism are twins. (Cf. Hegel's treatment of the relations between what he calls 'the first attitude of thought to objectivity' – pre-Kantian metaphysics – and the empiricist variant of the second attitude in the presentation of his system of logic in his *Encyclopaedia of the Philosophical Sciences* – 'Encyclopaedia Logic'/'Lesser Logic' – especially sections 37–39, though what is said here and what in the text to which this is a note are by no means the same.) It must be immediately added that the Slezak position has the inestimable merit of being intransigently materialist, so that it might be characterised, most generally, as a combination of an idealism (*qua* rationalism) in epistemology and a materialism (he might prefer the term 'realism') in ontology. It is not that Bloor 'goes too far' with his

'naturalistic' approach to knowledge, but rather that he does not go far enough: the 'loss of the object' that threatens in his account threatens the whole 'naturalistic' edifice.
22. I have presented some related considerations elsewhere – especially (1986), study 1, (1991), (1994), pp. 3–10, and (1995), and I refer the reader to these and to the literature cited there for expansions, here and there, of what I am unable to do more than at best briefly indicate in the present section.
23. Such signs may be called 'representations', so long as this term is taken as free of any implications of a one-one 'correspondence' between representation and what is represented and of what may be called 'mentalism'. On the second point cf. Toulmin (1972), pp. 192ff.
24. Cf. Plato, *Theaet.* 200d–201d: it is a necessary condition for a belief to constitute knowledge that it be supported by a *logos*.
25. To adopt Aristotle's terminology – via its Latin translations – in his analysis of change, a) is the 'material cause', and b) the 'efficient' cause respectively. Similarly, the considerations in the preceding paragraph (3) are about the 'formal cause' of knowledge. One reason for not using this terminology here is that the final member of the quartet, the 'final cause', is ambiguous as between a sense implying an intention or purpose on the one hand, and a purely functional sense on the other.
26. The question of the nature of knowledge has been thoroughly confused by the distinction, in effect introduced by Russell (though to be found in William James, *Principles of Psychology*, I, p. 221), between 'knowledge by acquaintance' and 'knowledge by description. The first expression is ambiguous as between two ideas, viz. (a) that of an absolutely direct form of knowledge, in principle at least without any element of discursiveness ('description'), and (b) that of a form of knowledge which, though at least potentially discursive ('descriptive'), has a distinctively direct connection with the causes that generate it. The position in the text to which this is a footnote is that (a) is an inadmissible idea, *all* knowledge being 'knowledge by description' in the sense of belonging, at least in principle, to the discursive domain. – It may be added that simple perceptual knowledge ('knowledge by acquaintance' in the sense accepted here) is fundamental insofar as it is the ultimate *causal* 'interface' between human beings and the natural world in which they find themselves but is not *cognitively* ultimate in the sense of received 'foundationalist' conceptions of knowledge.
27. This broadly defines the Aristotelian 'final cause' for everyday knowledge.
28. This is, of course, a fundamental part of Wittgenstein's account of language. Essentially the same point is approached from a somewhat different direction in Doyal & Harris (1983), taken up into their (1986), Ch. 7. (All this is of immediate relevance for the 'prison-house of language' thesis, but this implication cannot be pursued here.) It may be noted that this centrality of the social-practical dimension is fully consistent with the idea of innate capacities for language-learning of Chomskyan provenance.
29. On the whole subject of the social character of perception see especially Holzkamp (1986).
30. Cf. here Collingwood's account of 'absolute presuppositions' in his (1938), Toulmin's discussions in his (1963) of 'ideals of natural order', Popper (1982, pp. 31ff, 160ff).
31. See Pap's brilliant but neglected (1946) and the literature referred to there, especially the writings of V.F. Lenzen. – As an example from outside the physical sciences, the idea of natural selection may serve: originating in Darwin's substantive theory of evolution, it is now a general methodological tool. See, for instance, Edelman (1987) and other books by him. I have applied it in my (1983), Ch. 14 to the interpretation of Marx's theory of ideology.
32. On the significance of 'shared knowledge' see Gad Freudenthal's excellent (1984).
33. Cf. here, for instance, Gideon Freudenthal (1986, 1988), on the social origins of the 'analytic-synthetic' method, and related matters, and the literature cited there. See also the fine study, Gould (1994), explicitly pointed to the 'realism'/'relativism' controversy.
34. It is at this point that a more extensive treatment would have to take up the received distinction between 'context of discovery' and 'context of justification' (and the related

'externalist' versus 'internalist' controversies concerning the historiography of science). It must suffice here to say that there is a place for something like these distinctions though in a thoroughly *contextualised* fashion. The key point remains that briefly indicated in the main text to which this is a note, namely, that everything entering the discursive domain of science is subject to principles of validation accepted at a time, though these principles are themselves subject to appraisal as science develops. (For a comprehensive and subtle tracing of the development of probabilitistic methods – in part of social provenance – see Hacking 1975 and 1990.)

35. Cf. here especially the title piece in Althusser (1990).
36. See Suchting (1986, pp. 19ff), Baltas (1993), and the references given there.
37. Kitcher (1993, Chs. 6–8) also develops a means-end view of rationality according to which the latter is, most succinctly, just 'good cognitive design' (p. 182). He points out that a means-end view is at odds with much thinking about rationality, according to which the latter 'is taken to be constituted by a set of rules whose status is independent of their tendency to promote any ends' (p. 179, n. 3). It should be clear from the ensuing discussion in the main text that I am thoroughly sympathetic to his view that we should 'dissolve' rather than 'refine' the notion of rationality (p. 194 and Ch. 6 in general).
38. Here Braudel's distinction – see, e.g., (1980), p. 74 – between three sorts of temporal period (namely, those pertaining to 'events', 'conjunctures' and 'structures') may be useful.
39. On the historical origins of the idea of abstract rationality see Müller (1977).

REFERENCES

Althusser, L.: 1990, *Philosophy and the Spontaneous Philosophy of the Scientists and Other Essays*, Verso, London/New York.
Archer, M.: 1987, 'Resisting the Revival of Relativism', *International Sociology* 2, 219–223.
Baltas, A.: 1993, 'Physics as a Mode of Production', *Science in Context* 6, 569–616.
Barnes, B.: 1976, 'Natural Rationality: A Neglected Concept in the Social Sciences', *Philosophy of the Social Sciences* 6, 115–126.
Barnes, B.:1982, *T.S. Kuhn and Social Science*, Macmillan, London.
Barnes, B. & Bloor, D.: 1982, 'Relativism, Rationalism and the Sociology of Knowledge', in M. Hollis & S. Lukes (eds.), *Rationality and Relativism*, Basil Blackwell, Oxford, pp. 21–47.
Bloor, D.: 1976, *Knowledge and Social Imagery*, Routledge and Kegan Paul, London.
Bloor, D.: 1978, 'Polyhedra and the Abominations of Leviticus', *British Journal for the History of Science* 11, 243–272.
Bloor, D.: 1983, *Wittgenstein: A Social Theory of Knowledge*, Columbia University Press, New York.
Bloor, D.: 1984, 'The Strengths of the Strong Programme', in J.R. Brown (ed.), *Scientific Rationality: The Sociological Turn*, D. Reidel, Dordrecht/Boston/Lancaster, pp. 75–94.
Bloor, D.: 1988, 'Rationalism, Supernaturalism, and the Sociology of Knowledge', in I. Hronszky, M. Feher & B. Dajka (eds.), 1988, *Scientific Knowledge Socialized*, Kluwer Academic Publishers, Dordrecht, pp. 59–73.
Bloor, D.: 1991, 'Afterword:Attacks on the Strong Programme', in *Knowledge and Social Imagery* (2nd ed.), University of Chicago Press, Chicago, pp. 163–185.
Bloor, D.: 1992, 'Ordinary Human Inference as Material for the Sociology of Knowledge', *Social Studies of Science* 22, 129–139.
Bloor, D.: 1994, 'What Can the Sociologist of Knowledge Say About 2 + 2 = 4?', in P. Ernest (ed.), *Mathematics, Education and Philosophy: An International Perspective*, Falmer Press, Falmer, pp. 21–32.

Bloor, D.: (1996), 'The Question of Linguistic Idealism Revisited', in H. Sluga & D. Stern (eds.), *Cambridge Companion to Wittgenstein*, Cambridge University Press, Cambridge (forthcoming).
Braudel, F.: 1980, *On History*, University of Chicago Press, Chicago.
Brown, D.E.: 1991, *Human Universals*, Temple University Press, Philadelphia.
Bunge, M.: 1991, 'A Critical Examination of the New Sociology of Science, Part I', *Philosophy of the Social Sciences* 21, 524–560.
Bunge, M.: 1992 'A Critical Examination of the New Sociology of Science, Part II', *Philosophy of the Social Sciences* 22, 47–76.
Cohen, H.F.: 1994, *The Scientific Revolution. A Historiographical Inquiry*, University of Chicago Press, Chicago/London.
Collingwood, R.G.: 1940, *An Essay on Metaphysics*, Clarendon Press, Oxford.
Doyal, L. & Harris, R.: 1983, 'The Practical Foundations of Human Understanding', *New Left Review*, No. 139, May-June, pp. 59–78.
Doyal, L. & Harris, R.: 1986, *Empiricism, Explanation and Rationality. An Introduction to the Philosophy of the Social Sciences*, Routledge & Kegan Paul, London.
Edelman, G.M.: 1987, *Neural Darwinism. The Theory of Neuronal Group Selection*, Basic Books, New York.
Feyerabend, P.: 1975, *Against Method. Outline of an Anarchistic Theory of Knowledge*, New Left Books, London.
Feyerabend, P.: 1981, *Philosophical Papers*, 2 Vols., Cambridge University Press, Cambridge.
Forman, P.: 1971, 'Weimar Culture, Causality, and Quantum Theory, 1918–1927: Adaptation by German Physicists and Mathematicians to a Hostile Intellectual Environment', *Historical Studies in the Physical Sciences* 3, 1–116.
Forman, P.: 1978, 'The Reception of an Acausal Quantum Mechanics in Germany and Britain', in S. Mauskopf (ed.), *The Reception of Unconventional Science*, AAAS Selected Symposium, 24, Boulder, Colorado, pp. 11–50.
Freudenthal, Gad: 1979, 'How Strong is Dr. Bloor's "Strong Programme"?', *Studies in History and Philosophy of Science* 10, 67–83.
Freudenthal, Gad: 1984, 'The Role of Shared Knowledge in Science: The Failure of the Constructivist Programme in the Sociology of Science', *Social Studies of Science* 14, 285–295.
Freudenthal, Gideon: 1986, *Atom and Individual in the Age of Newton*, D. Reidel, Dordrecht.
Freudenthal, Gideon: 1988, 'Towards a Social History of Newtonian Mechanics. Boris Hessen and Henryk Grossmann Revisited', in I. Hronszky, M. Feher & B. Dajka (eds.), *Scientific Knowledge Socialized*, Kluwer Academic Publishers, Dordrecht, pp. 193–211.
Gellatly, A.: 1980, 'Logical Necessity and the Strong Programme for the Sociology of Knowledge', *Studies in History and Philosophy of Science* 11, 325–339.
Giere, R.N.: 1989, 'Computer Discovery and Human Interests', *Social Studies of Science* 19, 638–643.
Goodman, N.: 1983, *Fact, Fiction, and Forecast* (4th ed.), Harvard University Press, Cambridge, MA.
Gould, S.J.: 1994, 'Shields of Expectation – and Actuality', in *Eight Little Piggies. Reflections in Natural History*, Penguin, Harmondsworth, pp. 409–426.
Hacking, I.: 1975, *The Emergence of Probability*, Cambridge University Press, Cambridge.
Hacking, I.: 1990, *The Taming of Chance*, Cambridge University Press, Cambridge.
Hendry, J.: 1980, 'Weimar Culture and Quantum Causality', *History of Science* 18, 155–180.
Hesse, M.: 1974, *The Structure of Scientific Inference*, Macmillan, London.
Hesse, M.: 1980, 'The Strong Thesis in the Sociology of Science', in *Revolutions and Reconstructions in the Philosophy of Science*, Harvester, Brighton, pp. 29–60.

Holzkamp, K.: 1986, *Sinnliche Erkenntnis. Historischer Ursprung und gesellschaftlicher Funktion der Wahrnehmung* (5th ed.), Athenäum, Frankfurt/M..

James, W.: 1955, 'Pragmatism's Conception of Truth', In *Pragmatism* [1907], World Publishing Co., Cleveland/New York, pp. 131-153.

Jameson, F.: 1988, 'Marxism and Historicism', in *The Ideology of Theory. Essays 1971-1986*, University of Minnesota Press, Minneapolis, pp. 148-177.

Jennings, R. :1984, 'Truth, Rationality and the Sociology of Science', *British Journal for the Philosophy of Science* **35**, 201-211.

Jennings, R.: 1988, 'Alternative Mathematics and the Strong Programme: Reply to Triplett', *Inquiry* **11**, 93-101.

Kaufmann, W.: 1980, 'The Hegel Myth and Its Method', in *From Shakespeare to Existentialism: An Original Study*, Princeton University Press, Princeton, NJ, pp. 92-128.

Kitcher, P.: 1993, *The Advancement of Science. Science Without Legend, Objectivity Without Illusions*, Oxford University Press, New York/Oxford.

Lakatos, I.: 1976, *Proofs and Refutations. The Logic of Mathematical Discovery*, Cambridge University Press, Cambridge.

Lakatos, I.: 1978, *Philosophical Papers* 2 Vols., Cambridge University Press, Cambridge.

Latour, B. & Woolgar, S.: 1986, *Laboratory Life: The Construction of Scientific Facts* (2nd ed.), Princeton University Press, Princeton.

Laudan, L.: 1984, 'The Pseudo-Science of Science?', in Brown, J.R. (ed.), *Scientific Rationality: The Sociological Turn*, D. Reidel, Dordrecht/Boston/Lancaster, pp. 41-73.

Mackie, J.L.: 1974, *The Cement of the Universe. A Study of Causation*, Clarendon Press, Oxford.

Mannheim, K.: 1952, 'Historicism' (1924), in *Essays on the Sociology of Knowledge*, Oxford University Press, New York, pp. 84-133.

Meinecke, F.: 1972, *Historism. The Rise of a New Historical Outlook* (1959), Routledge & Kegan Paul, London.

Merton, R.K.: 1970 [1938], *Science, Technology and Society in Seventeenth Century England*, Howard Fertig, New York.

Merton, R.K.: 1957, 'The Sociology of Knowledge' (1945), in *Social Theory and Social Structure*, Free Press, Glencoe, Ill., pp. 456-488.

Müller, R.W.:1977, *Geld und Geist. Zur Entstehungsgeschichte von Identitätsbewußtsein und Rationalität seit der Antike*, Campus, Frankfurt/New York.

Musgrave, A.: 1976, 'Method or Madness?', in R. Cohen, P. Feyerabend & M. Wartofsky (eds.), *Essays in Memory of Imre Lakatos*, Reidel, Dordrecht, pp. 457-491.

Pap, A.: 1946, *The A Priori in Physical Theory*, King's Crown Press, New York.

Poincaré, H.: 1952, *Science and Hypothesis* [1902], Dover, New York.

Popper, K.R.: 1961, *The Poverty of Historicism* (corr. 2nd ed.), Routledge & Kegan Paul, London.

Popper, K.R.: 1962, *The Open Society and Its Enemies* (4th ed.), Routledge & Kegan Paul, London.

Popper, K.R.: 1982, *Quantum Theory and the Schism in Physics*, Hutchinson, London.

Putnam, H. : 1991, 'The "Corroboration" of Theories' (1974), in R. Boyd, P. Gasper & J.D. Trout (eds.), *The Philosophy of Science*, MIT Press, Cambridge, MA, pp. 121-137.

Restivo, S.: 1992, *Mathematics in Society and History. Sociological Inquiries*, Kluwer Academic Publishers, Dordrecht/Boston/London.

Shapin, S.: 1994, *A Social History of Truth:Civility and Science in 17th Century England*, University of Chicago Press, Chicago.

Slezak, P.: 1989, 'Scientific Discovery by Computer as Empirical Refutation of the Strong Programme', *Social Studies of Science* **19**, 563-600.

Slezak, P.: 1989a, 'Computers, Contents and Causes: Replies to My Respondents', *Social Studies of Science* **19**, 671-695.

Slezak, P.: 1991, 'Bloor's Bluff: Behaviourism and the Strong Programme', *International Studies in the Philosophy of Science* **5**, 241-256.

Slezak, P.: 1994, 'Sociology of Scientific Knowledge and Scientific Education: Part I', *Science & Education* **3**, 265-94.
Slezak, P.: 1994a, 'Sociology of Scientific Knowledge and Scientific Education, Part 2: Laboratory Life Under the Microscope', *Science & Education* **3**, 329-355.
Slezak, P.: 1994b, 'A Second Look at David Bloor's *Knowledge and Social Imagery*', *Philosophy of the Social Sciences* **24**, 336-361.
Slezak, P.: 1994c, 'The Social Construction of Social Constructionism', *Inquiry* **37**, 139-157.
Stehr, N. & Meja, V.: 1982, 'Zur gegenwärtigen Lage wissenssoziologischer Konzeptionen', in N. Stehr & V. Meja (eds.), *Der Streit um die Wissenssoziologie*, Vol. 2, Suhrkamp, Frankfurt/M., pp. 893-946.
Stove, D.C.: 1986, *The Rationality of Induction*, Clarendon Press, Oxford.
Suchting, W.: 1972, 'Marx, Popper, and "Historicism"', *Inquiry* **15**, 235-66.
Suchting, W.: 1983, *Marx. An Introduction*, Wheatsheaf, Brighton.
Suchting, W.: 1986, *Marx and Philosophy. Three Studies*, Macmillan, London.
Suchting, W.: 1991, 'On Some Unsettled Questions Touching the Character of Marxism, especially as Philosophy', *The Graduate Faculty Philosophy Journal*, New School for Social Research (New York), **14**, 139-207.
Suchting, W.: 1994, 'Notes on the Cultural Significance of the Sciences', *Science & Education* **3**, 1-56.
Suchting, W.: 1995, 'The Nature of Scientific Thought', *Science & Education* **4**, 1-22.
Toulmin, S.: 1963, *Foresight and Understanding. An Enquiry into the Aims of Science*, Harper and Row, New York/Evanston.
Toulmin, S.: 1972, *Human Understanding*, Vol. I, Clarendon Press, Oxford.
Triplett, T.: 1986, 'Relativism and the Sociology of Mathematics: Remarks on Bloor, Flew, and Frege', *Inquiry* **29**, 439-450.
Troeltsch, E.: 1922, *Der Historismus und seine Probleme. Erstes Buch: Das logische Problem der Geschichtsphilosophie*, Mohr, Tübingen.
Yearley, S.: 1982, 'The Relationship between Epistemological and Sociological Cognitive Interests', *Studies in History and Philosophy of Science* **13**, 253-288.

Educational Constructivism and Philosophy: Some References

MICHAEL R. MATTHEWS

School of Education Studies, University of New South Wales, Sydney 2052, Australia.

ABSTRACT: The research literature on educational constructivism is voluminous (see the Carmichael (1990) Pfundt & Duit (1994) and Driver et al. (1994b) bibliographies cited below). The research – in both the Piagetian and 'Alternative Conception' traditions – covers children's learning, cognitive development, curriculum development, classroom practices, teacher education, and much else. There is a further enormous literature on constructivism in philosophy of science (see Leplin (1984) and Churchland & Hooker (1985)), and on constructivism in the sociology of science (see Brown (1984), McMullin (1988, 1992). In turn these latter literatures overlap with the ocean of writing on postmodernist theory of knowledge and cognition (see Gross & Levitt (1994)). The following references relate mostly to educational constructivism, and then, with some exceptions, to articles that address epistemological and philosophical matters in science education. Even so it is not an exhaustive list, but hopefully it will be useful for teachers and researchers in the field. The author welcomes additions or omissions being brought to his attention.

Adey, P. S.: 1987, 'A Response to "Towards a Lakatosian Analysis of the Piagetian and Alternative Conceptions Research Programs"', *Science Education* **71**(1), 5-7.

Adey, P. S.: 1992, 'Alternative Constructs and Cognitive Development: Commonalities, Divergences and Possibilities for Evidence', *Research in Science Education* **22**, 1-10.

Ausubel, D. P.: 1964, 'Some Psychological Aspects of the Structure of Knowledge'. In S. Elam (ed.), *Education and the Structure of Knowledge*, Rand McNally, Chicago.

Ausubel, D. P.: 1985, 'Learning as Constructing Meaning'. In N. Entwistle (ed.), *New Directions in Educational Psychology I. Learning and Teaching*, The Falmer Press, London.

Benson, G. D.: 1989, 'Epistemology and Science Curriculum', *Journal of Curriculum Studies* **21**(4), 329-344.

Bereiter, C.: 1994, 'Constructivism, Socioculturalism, and Popper's World 3', *Educational Researcher* **23**(7), 21-23.

Bickhard, M. H.: 1993, 'On Why Constructivism Does Not Yield Relativism', *Journal of Experimental and Theoretical Artificial Intelligence* **5**, 275-284.

Bodner, G. M.: 1986, `Constructivism: A Theory of Knowledge', *Journal of Chemical Education* **63**(10), 873-878.

Brass, K. & Duke, M.: 1994, 'Primary Science in an Integrated Curriculum'. In P. Fensham, R. Gunstone & R. White (eds.) *The Content of Science: A Constructivist Approach to its Teaching and Learning*, Falmer Press, London, pp. 100-111.

Brown, J. R. (ed.): 1984, *Scientific Rationality: The Sociological Turn*, Reidel, Dordrecht.

Brown, J. R.: 1994, *Smoke and Mirrors: How Science Reflects Reality*, Routledge, New York.
Carmichael, P. et al.: 1990, *Research on Student's Conceptions in Science: A Bibliography*, Children's Learning in Science Project, University of Leeds, Leeds.
Churchland, P.M. & Hooker, C.A. (eds.): 1985, *Images of Science*, University of Chicago Press, Chicago.
Cobb, P. et al. 1991: 'Analogies from the Philosophy and Sociology of Science for Understanding Classroom Life', *Science Education* **75**(1), 23-44.
Cobb, P.: 1994, 'Where is the Mind? Constructivist and Sociocultural Perspectives on Mathematical Development', *Educational Researcher* **23**(7), 13-20.
Cobern, W. W.: 1993, 'Contextural Constructivism: The Impact of Culture on the Learning and Teaching of Science'. In K. Tobin (ed.) *Constructivist Perspectives on Science and Mathematics Education*, American Association for the Advancement of Science, Washington DC.
Confrey, J.: 1990, 'What Constructivism Implies for Teaching'. In R. Davis, C. Maher, & N. Noddings (eds.) *Constructivist Views on the Teaching and Learning of Mathematics*, National Council of Teachers of Mathematics, Reston, VA., pp.107-124.
Davis, R., Maher, C. & Noddings, N. (eds.): 1990, *Constructivist Views on the Teaching and Learning of Mathematics*, National Council of Teachers of Mathematics, Reston, VA.
Devitt, M.: 1991, *Realism & Truth*, 2nd ed., Basil Blackwell, Oxford.
Driver, R. & Bell, B.: 1986, `Students' Thinking and the Learning of Science: A Constructivist View', *School Science Review* **67**, 443-456.
Driver, R. & Easley, J.: 1978, `Pupils & Paradigms: A Review of Literature Related to Concept Development in Adolescent Science Students', *Studies in Science Education* **5**, 61-84.
Driver, R. & Oldham, V.: 1986, 'A Consructivist Approach to Curriculum Development in Science', *Studies in Science Education* **13**, 105-122.
Driver, R. et al.: 1994a, 'Constructing Scientific Knowledge in the Classroom', *Educational Researcher* **23**(7), 5-12.
Driver, R. et al.: 1994b, *Making Sense of Secondary Science*, Routledge, London.
Driver, R. et al.: 1994c, 'Working from Children's Ideas: Planning and Teaching a Chemistry Topic from a Constructivist Perspective'. In P. Fensham, R. Gunstone & R. White (eds.) *The Content of Science: A Constructivist Approach to its Teaching and Learning*, Falmer Press, London, pp. 201-220.
Duckworth, E.:1987, *The Having of Wonderful Ideas*, Teachers College Press, Columbia University, New York.
Duit, R.: 1993, 'Research on Students' Conceptions: Developments and Trends'. Paper presented at the Third International Conference on Misconceptions in Science and Mathematics Learning, Cornell University.
Duit, R.: 1993, `The Constructivist View: A Fashionable and Fruitful Paradigm for Science Education Research and Practice'. In L.P. Steffe (ed.) *Epistemological Foundations of Mathematical Experience*, Springer-Verlag, New York.

Ernest, P. (ed.): 1994, *Constructing Mathematical Knowledge: Epistemology and Mathematics Education*, The Falmer Press, London.

Fensham, P. J., Gunstone, R., & White, R. (eds.): 1994, *The Content of Science: A Constructivist Approach to its Teaching and Learning*, Falmer Press, London.

Fensham, P. J.: 1983, 'A Research Base for New Objectives in Science Teaching', *Science Education* **67**, 3-12.

Fensham, P. J.: 1992, 'Science and Technology'. In P.W. Jackson (ed.) *Handbook of Research on Curriculum*, Macmillan, New York, pp.789-829.

Fosnot, C. T.: 1993, 'Rethinking Science Education: A Defense of Piagetian Constructivism', *Journal of Research in Science Teaching* **30**(9), 1189-1201.

Fosnot, C.T. (ed.): 1996, *Constructivism: Theory, Perspectives, and Practice*, Teachers College Press, New York.

Garrison, J. W.: 1995, 'Deweyan Pragmatism and the Epistemology of Contemporary Social Constructivism', *American Educational Research Journal* **32**(4), 716-740.

Garrison, J.W.: 1997, 'An Alternative to von Glasersfeld's Subjectivism in Science Education: Deweyan Social Constructivism', *Science & Education* **6**(6),

Garrison, J.W. & Bentley, M.: 1989, 'Science Education, Conceptual Change, and Breaking with Everyday Experience', *Studies in Philosophy and Education* **10**(1), 19-36.

Geary, D. C.: 1995, 'Reflections of Evolution and Culture in Children's Cognition: Implications for Mathematical Development and Instruction', *American Psychologist* **50**(1), 24-37.

Geelan, D.R.: 1967, 'Epistemological Anachy and the Many Forms of Constructivism', *Science & Education* **6** (1-2), 15-28.

Gilbert, J. K. & Watts, D. M.: 1983, 'Concepts, Misconceptions & Alternative Conceptions: Changing Perspectives in Science Education', *Studies in Science Education* **10**, 61-98.

Gilbert, J.: 1993, 'Constructivism and Critical Theory'. In B. Bell (ed.), *I Know About LISP But How Do I Put It into Practice: Final Report of the Learning in Science Project (Teacher Development)*, Centre for Science and Mathematics Education Research, University of Waikato, Hamiliton.

Glasersfeld, E. von (ed.): 1991, *Radical Constructivism in Mathematics Education*, Reidel, Dordrecht.

Glasersfeld, E. von: 'Constructivism Reconstructed: A Reply to Suchting', *Science & Education* **1**(4), 379-384.

Glasersfeld, E. von: 1987, *Construction of Knowledge*, Intersystems Publications, Salinas CA.

Glasersfeld, E. von: 1989, `Cognition, Construction of Knowledge, and Teaching', *Synthese* **80**(1), 121-140.

Glasersfeld, E. von: 1990, 'An Exposition of Constructivism: Why Some Like It Hot'. In Davis, R., Maher, C. & Noddings, N. (eds.), *Constructivist Views on the Teaching and Learning of Mathematics*, National Council of Teachers of Mathematics, Reston, VA., pp.19-30.

Glasersfeld, E. von: 1992, 'Questions and Answers About Radical Constructivism'. In M. K. Pearsall (ed.) *Scope, Sequence, and Coordination of Secondary School Science, Vol.11, Relevant Research*, NSTA, Washington DC., pps.169-182.

Glasersfeld, E. von: 1995, *Radical Constructivism. A Way of Knowing and Learning*, The Falmer Press, London.

Goldin, G.: 1990, 'Epistemology, Constructivism, and Discovery Learning in Mathematics'. In Davis, R., Maher, C. & Noddings, N. (eds.) *Constructivist Views on the Teaching and Learning of Mathematics*, National Council of Teachers of Mathematics, Reston, VA.

Good, R., Wandersee, J. & St. Julien, J.: 1993, 'Cautionary Notes on the Appeal of the New "Ism" (Constructivism) in Science Education'. In K. Tobin (ed.) *Constructivism in Science and Mathematics Education*, AAAS, Washington DC, pp. 71-90.

Gross, P. R. & Levitt, N.: 1994, *Higher Superstition: The Academic Left and Its Quarrels with Science*, Johns Hopkins University Press, Baltimore.

Hardy, M.D. & Taylor, P.C.: 1997, ' Von Glasersfeld's Radical Constructivism: A Critical Review', *Science & Education* **6** (1-2), 135-150.

Hawkins, D.: 1994, 'Constructivism: Some History'. In P. Fensham, R. Gunstone & R. White (eds.) *The Content of Science: A Constructivist Approach to its Teaching and Learning*, Falmer Press, London, pp. 9-13.

Hendry, G. D. & King, R. C.: 1994, 'On Theory of Learning and Knowledge: Educational Implications of Advances in Neuroscience', *Science Education* **78**(3), 223-253.

Jonassen, D. H.: 1991, 'Objectivism versus Constructivism: Do We Need a New Philosophical Paradigm?', *Educational Technology, Research and Development* **39**(3), 5-14.

Kelly, G., Carlsen, W., & Cunningham, C.: 1993, 'Science Education in Sociocultural Context: Perspectives from the Sociology of Science', *Science Education* **77**(2), 207-220.

Kelly, G.J.: 1997, 'Research Traditions in Comparative Context: A Philosophical Challenge to Radical Constructivism', *Science Education* **81**(3), 355-375.

Kilpatrick, J.: 1987, `What Constructivism Might Be in Mathematics Education'. In J.C. Bergeron, N. Herscovics, & C. Keiran (eds.) *Psychology of Mathematics Education*, Proceedings of the Eleventh International Conference, Montreal, pp.3-27.

Kitchener, R. F.: 1985, 'A Bibliography of Philosophical Work on Piaget', *Synthese* **65**(1), 139-151.

Kitchener, R. F.: 1986, *Piaget's Theory of Knowledge: Genetic Epistemology and Scientific Reason*, Yale University Press, New Haven.

Kitchener, R. F.: 1993, 'Piaget's Epistemic Subject and Science Education: Epistemological Versus Psychological Issues', *Science & Education* **2**(2), 137-148.

Konold, C. & Johnson, D. K.: 1991, 'Philosophical and Psychological Aspects of Constructivism'. In L.P. Steffe (ed.) *Epistemological Foundations of Mathematical Experience*, Springer-Verlag, New York.

Lave, J.: 1988, *Cognition In Practice: Mind, Mathematics And Culture In Everyday Life*, Cambridge University Press, New York.

Leplin, J. (ed.): 1984, *Scientific Realism*, University of California Press, Berkeley.

Mahoney, M. J.: 1988, 'Constructivist Approaches in Educational Research', *Review of Educational Research* **47**, 651-693.

Matthews, M. R.: 1992a, 'Constructivism and Empiricism: An Incomplete Divorce', *Research in Science Education* **22**, 299-307.
Matthews, M. R.: 1992b, 'Constructivism and the Empiricist Legacy'. In M.K. Pearsall (ed.) *Scope, Sequence and Coordination of Secondary School Science: Relevant Research*, National Science Teachers Association, Washington, DC, pp.183-196.
Matthews, M. R.: 1993, 'Constructivism and Science Education: Some Epistemological Problems', *Journal of Science Education and Technology* **2**(1), 359-370.
Matthews, M. R.: 1994, *Science Teaching: The Role of History and Philosophy of Science*, Routledge, New York.
Matthews, M.R.: 1995, *Challenging New Zealand Science Education*, Dunmore Press, Palmerston North.
Matthews, P.: 1997, 'Problems with Piagetian Constructivism', *Science & Education* **6**(1-2), 105-119.
McMullin, E. (ed.): 1988, *Construction and Constraint*, University of Notre Dame Press, Notre Dame, IN.
McMullin, E. (ed.): 1992, *The Social Dimensions of Science*, University of Notre Dame Press, Notre Dame, IN.
Millar, R. & Driver, R.: 1987, 'Beyond Processes', *Studies in Science Education* **14**, 33-62.
Millar, R.: 1989, 'Constructive Criticisms', *International Journal of Science Education* **11**, 587-596.
Nadeau, R. & Destautels, J.: 1984, *Epistemology and the Teaching of Science*, Science Council of Canada, Ottawa.
Novak, J. D.: 1977, 'An Alternative to Piagetian Psychology for Science and Mathematics Education', *Science Education* **61**(4), 453-477.
Novak, J. D.: 1987, 'Human Constructivism: Toward a Unity of Psychological & Epistemological Meaning Making'. In J. D. Novak (ed.), 1987, *Proceedings of the Second International Seminar on Misconceptions & Educational Strategies in Science & Mathematics* vol I, pp.349-360.
Nussbaum, J.: 1989, 'Classroom Conceptual Change: Philosophical Perspectives', *International Journal of Science Education* **11**(5), 530-540.
Ogborn, J.: 1997, 'Constructivist Metaphors of Science Learning', *Science & Education* **6**(1-2), 121-133.
O'Loughlin, M.: 1992, 'Rethinking Science Education: Beyond Piagetian Constructivism Toward a Sociocultural Model of Teaching and Learning', *Journal of Research in Science Teaching* **29**(8), 791-820.
Osborne, J.: 1996, 'Beyond Constructivism', *Science Education* **80**(1), 53-82.
Osborne, R. & Wittrock, M.: 1985, 'The Generative Learning Model and its Implications for Science Education', *Studies in Science Education* **12**, 59-87.
Otte, M.: 1994, 'Is Radical Constructivism Coherent?'. In P. Ernest (ed.), *Constructing Mathematical Knowledge: Epistemology and Mathematics Education*, The Falmer Press, London.
Pfundt, H. & Duit, R.: 1994, *Bibliography of Students' Alternative Frameworks & Science Education*, 4th Edit., Institute for Science Education, University of Kiel.
Phillips, D.C.: 1995, 'The Good, the Bad and the Ugly: The Many Faces of Constructivism', *Educational Researcher* **24**(7), 5-12.

Piaget, J.: 1970, *Genetic Epistemology*, Columbia University Press, New York.
Piaget, J.: 1972, *Psychology and Epistemology: Towards a Theory of Knowledge*, Penguin, Harmondsworth.
Piattelli-Palmarini, M.(ed.): 1980, *Language and Learning: The Debate Between Jean Piaget and Noam Chomsky*, Routledge & Kegan Paul, London.
Piattelli-Palmarini, M.: 1994, 'Ever Since Language and Learning: Afterthoughts on the Piaget-Chomsky Debate', *Cognition* **50**, 315-346.
Pope, M. L. & Keen, T. R.: 1981, *Personal Construct Psychology and Education*, Academic Press, London.
Pope, M.L. & Gilbert, J.: 1983, 'Personal Experience and the Construction of Knowledge in Science', *Science Education* **67**(2), 193-203.
Posner, G. et al.: 1982, 'Accommodation of a Scientific Conception : Toward a Theory of conceptual Change', *Science Education* **66**(2), 211-227.
Richards, J. & Glasersfeld, von E.: 1979, 'The Control of Perception and the Construction of Reality', *Dialectica* **33**(1), 37-58.
Roth, M.-W.: 1995, *Authentic School Science: Knowing and Learning in Open-Inquiry Science Laboratories*, Kluwer Academic Publishers, Dordrecht.
Roth, M-W., & Roychoudhury, A.: 1994, 'Physics Students' Epistemologies and Views about Knowing and Learning', *Journal of Research in Science Teaching* **31**(1), 5-30.
Rowell, J. A.: 1989, 'Piagetian Epistemology: Equilibration and the Teaching of Science', *Synthese* **80**(1), 141-162.
Samarapungavan, A.: 1992, 'Children's Judgements in Theory-Choice Tasks: Scientific Rationality in Childhood', *Cognition* **45**, 1-32
Slezak, P.: 1994a, 'Sociology of Scientific Knowledge and Science Education: Part I', *Science & Education* **3**(3), 265-294.
Slezak, P.: 1994b, 'Sociology of Scientific Knowledge and Science Education: Part II', *Science & Education* **3**(4), 329-356.
Solomon, J.: 1994, 'The Rise and Fall of Constructivism', *Studies in Science Education* **23**, 1-19.
Steffe, L. & Gale, J. (eds.): 1995, *Constructivism in Education*, Lawrence Erlbaum Associates, Hillsdale, NJ.
Steffe, L. (ed.): 1991, *Epistemological Foundations of Mathematical Experience*, Springer-Verlag, New York.
Strike, K. A. & Posner, G.J.: 1992, 'A Revisionist Theory of Conceptual Change'. In R. Duschl & R. Hamilton (eds.) *Philosophy of Science, Cognitive Psychology, and Educational Theory and Practice*, State University of New York Press, Albany, NY., pp.147-176.
Strike, K. A.: 1987, 'Towards a Coherent Constructivism'. In J. D. Novak (ed.) *Misconceptions & Educational Strategies*, Education Department, Cornell University, vol.I. pp.481-489.
Suchting, W. A.: 1992, 'Constructivism Deconstructed', *Science & Education* **1**(3), 223-254.
Tobin, K. (ed.): 1993, *The Practice of Constructivism in Science and Mathematics Education*, AAAS Press, Washington DC.

Tobin, K.: 1991, 'Constructivist Perspectives on Research in Science Education', paper presented at the annual meeting of the National Association for Research in Science Teaching, Lake Geneva, Wisconsin.

Vosnaidou, S. & Brewer, W. F.: 1987, 'Theories of Knowledge Restructuring in Development', *Review of Educational Research* **37**(1), 51-67.

Vygotsky, L. S.: 1962, *Thought and Language*, MIT Press, Cambridge.

Vygotsky, L. S.: 1978, *Mind in Society: The Development of Higher Psychological Processes*, Harvard University Press, Cambridge.

Watts, D. M. & Bentley, D.: 1991, 'Constructivism in the Curriculum. Can We Close the Gap between the Strong Theoretical Version and the Weak Version of Theory in Action?', *The Curriculum Journal* **2**(2), 171-182.

Watts, D.M. & Bentley, D.: 1994, 'Humanizing and Feminizing School Science: Reviving Anthropomorphic and Animistic Thinking in Constructivist Science Education', *International Journal of Science Education* **16**(1), 83-98.

Watts, D.M.: 1994, 'Constructivism, Re-constructivism and Task-orientated Problem-solving'. In P. Fensham, R. Gunstone & R. White (eds.) *The Content of Science: A Constructivist Approach to its Teaching and Learning*, Falmer Press, London, pp. 39-58.

Wheatley, G.H.: 1991, 'Constructivist Perspectives on Science and Mathematics Learning', *Science Education* **75**(1), 9-22.

Yeany, R.H.: 1991, 'A Unifying Theme in Science Education?', *NARST News* **33**(2), 1-3.

NOTES ON THE CONTRIBUTORS

Mark Bickhard is Henry R. Luce Professor of Cognitive Robotics and the Philosophy of Knowledge at Lehigh University. He holds degrees in mathematics, statistics and human development, all from the University of Chicago. His research interests span the issues of what persons – natural or artificial – are, and how they know and learn about their worlds. He has published extensively in the fields of cognitive learning theory, developmental psychology, and artificial intelligence.

Ernst von Glasersfeld was a professor of psychology at the University of Georgia. He is attached to the Scientific Reasoning Research Institute at the University of Massachusetts, Amherst. He has published many papers on linguistics, Piagetian psychology, mathematics pedagogy, and constructivist theory. A number of these have been published in his *The Construction of Knowledge* (Intersystems Publications 1987). He has recently published *Radical Constructivism: A Way of Knowing* (Falmer Press, 1995), and edited *Radical Constructivism in Mathematics Education* (Kluwer 1991).

Richard E. Grandy is a professor in the Department of Philosophy and the Program in Cognitive Sciences at Rice University. He received a B.S. in mathematics from the University of Pittsburgh in 1963, and an M.A. (1965) and Ph.D. (1968) in the History and Philosophy of Science from Princeton University. His previous appointments were at Princeton University and the University of North Carolina, Chapel Hill. His current interests include the significance of philosophy of science and the cognitive sciences for science education.

Helge Kragh is a professor at the Institute for the History of Exact Sciences, Aarhus University, Denmark. He studied physics and chemistry at the University of Copenhagen and received his Ph.D. in 1981 from Roskilde University Centre, Denmark. His research interests cover the history of modern physics, historiography of science, and history of 19th-century chemistry. Among his recent publications are *An Introduction to the Historiography of Science* (1987) and *Dirac: A Scientific Biography* (1990), both published by Cambridge University Press.

Michael R. Matthews is a senior lecturer in the School of Education Studies at The University of New South Wales. He has degrees from the University of Sydney in science, philosophy, psychology, history and philosophy of science, and education. He has taught in high school, at Sydney Teachers' College, and was the Foundation Professor of Science Education at the University of Auckland (1992-93). He has published in philosophy of education, philosophy of science, and science education. His recent books include *Challenging New Zealand Science Education* (Dunmore Press, 1995), and *Science Teaching:*

The Role of History and Philosophy of Science (Routledge, 1994), and *Time for Science Education* (Plenum, 1998). He has edited *The Scientific Background to Modern Philosophy* (Hackett Publishing Company, 1989), and *History, Philosophy and Science Teaching: Selected Readings* (OISE Press/Teachers College Press, 1991).

Robert Nola is an associate professor of philosophy at the University of Auckland. He has degrees in science and philosophy. He has published in philosophy of science, 19th-century philosophy, and philosophy and science education. He is the editor of *Relativism and Realism in Science* (Kluwer Academic Publishers, 1988). His research areas are epistemology, metaphysics, and philosophy of science, especially issues to do with realism and relativism. He has been a Visiting Fellow at the Pittsburgh University Centre for Philosophy of Science.

Denis C. Phillips is professor of education and, by courtesy, professor of philosophy at Stanford University, where he also serves as the Associate Dean in the School of Education. He was born and educated in Australia where he graduated in science and taught high school science before completing his Ph.D. and lecturing at Monash University in education. He moved to Stanford in 1974. His professional interests encompass philosophy of education, philosophy of social science, social science and educational research methodology, and history of nineteenth and twentieth century thought (especially the interrelationships between philosophy, biology and psychology). His most recent books are *Philosophy, Science, and Social Inquiry* and *The Social Scientist's Bestiary* (both published by Pergamon Press, Oxford).

Peter Slezak is a senior lecturer in the School of Science and Technology Studies at the University of New South Wales. He graduated in sociology, and completed his Ph.D. in philosophy at Columbia University. He has published in philosophy of science, cognitive science and theoretical psychology. He is researching topics in mental imagery, and the theoretical foundations of sociology of science.

Wallis Suchting was a reader in the Department of General Philosophy at Sydney University. Sadly, he took his life in January 1997. He completed a B.A. in philosophy at the University of Queensland, taught high school history for a number of years, and completed in 1971 a Ph.D. in philosophy of science at Melbourne University. He had published extensively, especially in his main fields of interest, namely, history of philosophy, philosophy of science, and Marxism. He was co-translator and co-editor of Hegel's *Encyclopedia Logic* (Hackett Publishing Company 1991). Other books include *Marx: An Introduction* (Harvester Press, 1983), and *Marx and Philosophy* (Macmillan, 1986).

NAME INDEX

American Association for the Advancement of Science (AAAS)   x, 2
Anderson, C.D. 130
Apple, M. 6
Aristotle 211
Aronowitz, S. 6
Bacon, F. 208
Baldwin, M. 12
Barnes, B. 141, 144, 145, 166, 203
Berkeley, G. 69, 80, 97
Berlin, I. 37
Blondlot, R. 134, 160
Bloor, D. 140, 141, 146-150, 161-162, 165-174, 190-203
Braudrillard, J. 183, 184
Bronowski, J. 181, 184
Bruner, J.S. 109
Bunge, M. xi, 161
Burt, C. 164
Cartwright, N. 117
Ceccato, S. 27
Cherryholmes, C. 141
Chomsky, N. 183, 1984
Collingwood, R.G. 211
Collins, H.M. 129, 131, 132, 135, 141, 145, 163-164, 171, 177, 181
Cromer, A. xi
Dana, T.M. 56
Darwin, C. 15, 103
Davis, N.T. 56
Devitt, M. xi, 1
Dingler, H. 85
Douglas, M. 141
Douglas, M. 200
Driver, R. 8, 9, 56-57,
Duit, R. 2
Durkheim, E. 141, 179
Einstein, A. 48, 49, 181
Fensham, P. 2,
Feyerabend, P.K. 127

Feynman, R.   126
Fish, S.   141, 156
Fleck, L.   12, 27
Forman, P.   175
Foucault, M.   141
Fox-Keller, E.   141, 142
Fraassen, B. van   3, 32
Frankfort School   127
Franklin, A.   131
Freudenthal, G. 208, 211
Freyberg, P.   4
Fuller, S.   142, 143
Gadamer, H.-G.   105
Galileo, G.   82-83
Galison, P.   129
Giere, R.   118
Gieryn, T.   166
Giroux, H.   6
Glasersfeld, E. von   42, 45-53, 55, 61-85, 108, 139
Gross, P.   xi, 1, 128
Guthrie, W.K.C.   182
Hacking, I.   xi
Hankinson-Nelson, L.   142
Harding, S.   142
Hawking, S.   126
Hegel, W.G   209
Heidegger, M.   105
Hirst, P.H.   143
Hitler, A.   182
Hobbes, T.   80-82
Hume, D.   69, 75, 97, 190
International History, Philosophy and Science Teaching Group (IHPSTG)
    ix
James, W.   209
Kamerlingh-Onnes, H.   130
Kant, I.   20, 40-42, 75, 96, 104
Kitcher, P.   212
Kornblith, H.   155
Kuhn, T.S.   xii, 3, 11, 32, 127-128, 141, 142, 144
Labinger, J.   128
Lakatos, I.   154, 200, 210
Latour, B.   141, 151-152, 160

NAME INDEX

Laudan, L.   xi, 140, 154, 162, 168
Leeds University   2, 8
Levitt, N.   xi, 1
Locke, J.   7, 97, 139
Longino, H.   118-119, 142
Marx, K.   141
Mauss, M.   179
McMullin, E.   xi
Megill, A.   117
Merton, T.   164, 166, 190
Mill, J.S.   7, 196
Monash University   2, 9
Montaigne, M. de   7
Mulkay, M.   141
National Association for Research in Science Teaching (NARST)   1, 6
Needham, R.   179-180
Newton-Smith, W.H.   144
Norris, C.   183, 184
Orwell, G.   183
Osborne, R.   4
Pap, A.   211
Peirce, C.S.   102-103
Piaget, J.   2, 15-19, 26, 42-45, 67, 106-107, 139
Pickering, A.   129, 165
Pinch, T.   132, 163-164, 169, 181
Plato   34-37
Plowden Report   6
Popper, K.R.   41-42, 55, 107, 162
Posner, G.   2
Protagoras
Reichenbach, H.   41
Röntgen, W.C.   130, 160
Rorty, R.   14, 24, 140, 141, 143
Russell, B.   7, 159, 181
Schwab, J.J.   143
Science Education Portfolio Instruction and Assessment (SEIPA)
Shapin, S.   131, 135, 142, 173-174
Siegel, H.   183
Sismondo, S.   142
Slezak, P.   132, 134, 136, 152, 155, 189-203
Snow, C.P.   182
Socrates   7, 34-37,

Sokal, D.　　　xi, 156
Solomon, J.　　8, 9
Stove, D.　　　xi, 162
Strike, K.　　　2
Suchting, W.A.　　　93-98
Thorndike, E.　18
Tobin, K.　　　6, 31
Toulmin, S.E.　211
Uexküll, J. von　　　12
Vgotsky, L.S.　139
Vico, G.　　　12-15, 37-40, 70-74, 79-82
Weinberg, S.　128
White, R.　　　8
Wittgenstein, L.　　　63, 78, 202
Wolpert, L.　xi, 128
Woolgar, S.　141-143, 146, 151-153, 160

SUBJECT INDEX

Abstraction theories of concept formation   77-78
Accommodation, Piagetian   16
Adaptation   15, 26, 94
Anti-science movements   126-128
Aristotelian epistemology   83, 211n
Artificial intelligence   171
Assimilation, Piagetian   16, 44
Behaviourism   161, 170-171
Berkeley's epistemology   39, 69, 81
Bloor's theory of science (see also Edinburgh 'Strong Programme')   146-149, 154-155, 165-181, 192-204
British empiricists   48
Chomsky and the cultural role of intellectuals   183-184
Cognitive constructivism   114-115
Commonsense reasoning and knowledge   45, 83, 205
Construction in knowledge acquisition   42-45, 66, 69-71, 118
Content of science, constructivist teaching of   7-14
Conventionalism   197-199, 201-203
Correspondence theory of truth   14, 46-47, 101, 195-197
Creation science   160
Critical Theory   6
Deconstruction   93, 165
Discovery, scientific   130
Discovery/justification distinction   41
Duhem-Quine thesis (see Underdetermination of theory)
Durkheim's sociology of culture   179-181
Edinburgh 'Strong Programme' in sociology of scientific knowledge   3, 127-136, 144-156, 162-181, 192-204
    core theses   167
Empiricism (see also Berkeley, British empiricists)   68, 84, 97
Epistemic constructivism   117-119
Erlangen School   85n
Ethics   1, 6, 134, 163-165
Euclidean geometry   74
Evolutionary epistemology   107
Evolutionary theory   15-16, 211n

Experience, role in knowledge   48-49, 68-69
Fallibility, scientific   131
Feyerabendian epistemology   127
Galilean physics and methodology   64, 82-83, 88n, 203
Glasersfeld's von, epistemology   45-53, 61-85, 93-98, 108
Hegel   162
History of constructivism   13-15
History of science in science teaching   120
Hobbe's epistemology   80-82
Hypothetico-deductive method   41
Idealism   100-101, 104-105, 198-199
Indoctrination   159-160
Infant learning   17-20, 75-76, 96, 109
Influence of constructivism   1, 2, 125ff
Instrumental reason   127, 181-182
Instrumentalist theory of science (see also Adaptation, Viability)   12, 14, 48, 51
Interests and science   176-177
Kantian epistemology   20, 40-41, 104
Kuhnian theory of science   11, 32, 127, 144
Laboratories, school   133
Laboratory Life   151-152
Lakatosian theory of science   154, 200
Language, acquisition   21-24, 52, 77-79
    and ideology   183
Learning theory   2
    contrast with theory of science   55
Leeds University Research Group   8-14, 56-57
Logical positivism   39
Marxist epistemology   141
Meaning, theory of   200-203
Mertonian, norms of science   163-165
Metaphysical constructivism   114
Misconceptions   116, 133-134
Motivation   27
Naturalised epistemology   142
New Zealand science edcation   4ff
Non-transferability of knowledge (see also Transmission theory)   5, 22
N-rays   134, 160
Objectivism   31, 117-118
    contrast with certainty   64
Operationalism   48

SUBJECT INDEX

Pedagogy/Theory distinction   3, 7, 121
Pendulum experiments   133
Phenomenalism   48-49, 54
Piaget's constructivism   12, 15-19, 42-45, 96, 106-107
Platonic (and Socratic) epistemology   34-37
    and theories of pedagogy   56
Plowden Report   6
Popperian epistemology   41-42
Positivism   135, 210n
Postmodernism   1, 140-141, 150, 183-185
Pragmatism   14, 15, 102-104
Presocratic philosophy   159
Production theory of knowledge   89n, 206
Progress, scientific   130
Progressive education   6
Pseudoscience   126, 160
Psychological constructivists   139
Radical constructivism (see Glasersfeld's von, epistemology)
Rationalism   101-102, 183, 207, 210n, 212n
Realism   31-33, 100, 114, 131
Reflexivity   129
Relativism   129, 133, 145, 160, 193-203
Romanticism   126
Russell, Bertrand and education   181
Scaffolding, cognitive   109-110
Scheme theory   16-19
Science wars   128, 140, 156, 161-162
Scientific literacy   126, 182
SEIPA (Science Education Portfolio Instruction and Assessment)   113
Sense-making   5
Sensory pathways   14, 205, 208n
Skepticism (see also Relativism)   100
Social constructivism   19-21, 50, 57, 74-76, 118-119, 128ff, 140-156, 159-185,
    172-174
Social determination of science   169-177, 206
Sociology of science   166
Sociology of Scientific Knowledge (see Edinburgh Strong Programme)
Strong programme in sociology of scientific knowledge (see Edinburgh Strong
    Programme)
Transmission theory of learning   10
Two-cultures divide   125
Underdetermination of theory   177-179

Varities of constructivism   3, 99, 104-109, 113-116, 139
Viability (see also Instrumentalism)   15-16, 67
Vico's epistemology   13-15, 37-40, 70-72, 79-84, 94-95,
Vygotsky's learning theory   109
Wittgenstein's theories of language and knowledge   78, 86n, 143, 199, 202, 211n
Worldviews   6

DATE DUE

PRINTED IN U.S.A.